AF577122

PRINCIPLES AND PRACTICES OF SEED STORAGE

Oren L. Justice and
Louis N. Bass

CASTLE HOUSE PUBLICATIONS LTD.

ISBN 0 7194 0022 8
Castle House Publications Ltd 1979

Printed and bound in Great Britain at
William Clowes & Sons Limited, Beccles and London

CONTENTS

PRINCIPLES AND PRACTICES OF SEED STORAGE

By OREN L. JUSTICE[1] and LOUIS N. BASS[2]

INTRODUCTION

Purpose and Significance of Safe Seed Storage

The principal purpose of storing seeds of economic plants is to preserve planting stocks from one season until the next. Prehistoric man learned the necessity of this practice and developed methods of storing small quantities of seeds for his future use. As agriculture developed, man expanded his knowledge regarding both the requirements of seed for maintenance of viability and methods of providing suitable storage conditions. In 1832, Aug. Pyr. de Candolle of France included a chapter on seed preservation in his book "Physiologie Végétale." He pointed out that the vitality of seeds would be prolonged if stored under conditions to protect them from heat, moisture, and oxygen. About the same time, other authors suggested the use of tar-coated wooden boxes and iron tanks capable of being sealed with stopcocks for drawing off the seeds as desired.

Although the storage of seed stocks for planting the following season remains the most important reason for storing seeds, farmers and seedsmen have found it advantageous to carry over seeds for 2 or more years. This practice results in accumulating supplies of desired genetic stocks for use in years following periods of low production. Many kinds of seeds—mostly vegetable, flower, and forage seeds—move rather freely in world commerce. Many of these seed lots are not used the first year after production.

With increased knowledge and technology of plant genetics and plant breeding, the necessity for longtime storage of small quantities of the various cultivars becomes apparent. Facilities for storing genetic stocks now exist at Fort Collins, Colo., Hiratsuka, Japan, Braunschweig, Germany, Bari, Italy, Izmir, Turkey, and other locations. Research workers can obtain genetic stocks considered useful in their breeding programs from these facilities. The principles to be discussed here

[1] Formerly Agricultural Marketing Research Institute, Northeastern Region, Agricultural Research Service. Retired June 1973.

[2] National Seed Storage Laboratory, Western Region, Agricultural Research Service.

apply equally to all seed storage. Longtime storage of genetic stocks is not the principal objective of this handbook.

No statistics are available indicating either the actual losses sustained from substandard storage conditions or the savings effected by optimum storage conditions. However, the importance of suitable storage becomes apparent when one considers the alternatives. Failure to use available information about seed storage could greatly handicap the Nation's agricultural program as follows: (1) Agriculture in the warm, humid, subtropical, and tropical regions would be less efficient, (2) the plant breeder would be greatly handicapped by not maintaining genetic stocks, and (3) international commerce in seeds would be only a small fraction of its present economic value.

The Seed, a Fragile, Living Organism

In his poem "A Package of Seeds," Edgar A. Guest expressed very well the theme of this section. When Guest had purchased packages of seeds, he had thought of them only as *seeds.* But on one occasion when he made such a purchase, it flashed through his mind that he had purchased "a miracle of life," a dime's worth of power that no man can create and a dime's worth of mystery, destiny, and fate. In working with seeds, especially in harvesting, cleaning, handling, storage, and transportation, it is essential to keep in mind at all times that inside the seed is a dormant miniature plant awaiting the opportunity to continue its growth.

Basically the seed is made up of the embryo or miniature plant, the endosperm and other food reserves, and protective coverings consisting of the seedcoat and, in some cases, accessory structures (fig. 1). The embryo is more protected in the resting seed of corn than in bean. In corn and other grains the major food reserves are stored as endosperm, whereas in bean they are stored in the two cotyledons or seed leaves, which also serve as photosynthetic organs for the young plant. Other structures affecting seed storage are the seedcoat for both corn and bean and the hilum and micropyle of bean, which may permit the entry or exit of moisture. Under some conditions the hilum may become soft or damaged, allowing fungi to enter the seed.

The planting units in several plant families are fruits, botanically speaking, rather than true seeds. In the grass family the seed or fruit is called a grain or caryopsis, in the buckwheat and sunflower families an achene, and in the mint family a nutlet. Some kinds of seeds possess additional structures, such as glumes, bracts, spines, and hairs, which aid in protecting the seed from injury, birds, and rodents. In this publication, all these structures that normally serve as the planting unit are referred to as seeds.

The orientation of the embryo within the seed and the nature of the protective coverings are important attributes of any seed that is sub-

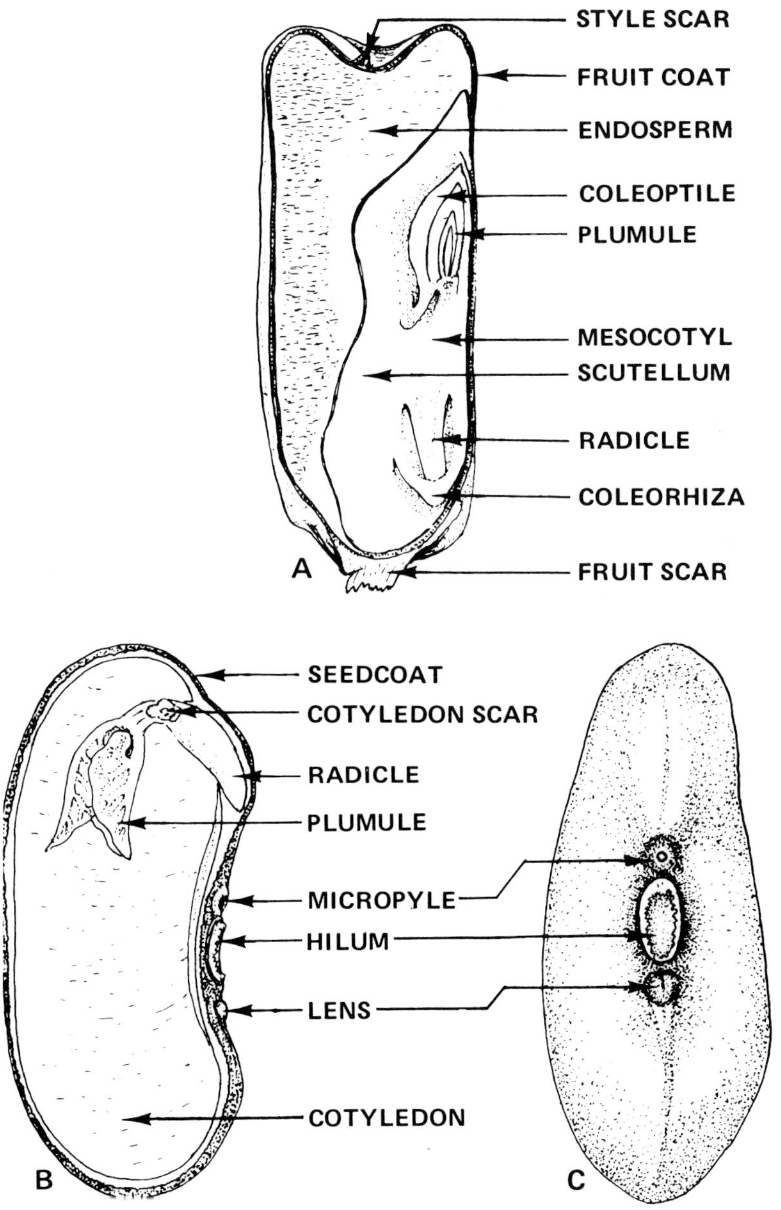

FIGURE 1.—External and internal structures of corn (*A*) and bean (*B* and *C*) seeds.

jected to the numerous mechanical and handling processes from harvest to planting. For example, a grass seed has its embryonic root tip near the base of the grain. When the glumes are removed during harvesting, as in rye, timothy, and wheat, the root tip is only protected by a thin membrane, which is very vulnerable to physical damage and attack by storage molds. However, seeds of many grasses, such as barley, oats, rice, and some cultivars of sorghum, are much better protected since the enclosing glumes usually remain attached.

The relatively large seeds of bean and soybean have only a moderately thick seedcoat. The plumule, or embryonic stem, is fairly well developed in the resting seed and lies between two cotyledons, or seed leaves. Also, the radicle, or embryonic root, has practically no protection except that provided by the seedcoat. Thus, a bean seed is unusually vulnerable to breakage, especially when dry and roughly treated. The radicle, plumule, or cotyledons can be damaged. Velvetbean, in the same plant family as bean and of similar structure, is less vulnerable to damage from harvesting, handling, and processing because of a slightly different orientation of the plumule and radicle.

From Harvest to Sowing

Beginning with harvest, seed lots usually pass through a series of processes necessary for immediate or future use. They include harvesting, drying, cleaning, grading or sizing, transporting, and storing. Not all commercial seed lots are subjected to all these processes, but the manner in which seeds are passed through each process can affect their storage potential.

Each seed faces an uncertain future. It may be harvested too early, too late, too wet, or too dry. The seed's future can be affected by being harvested so immature that it cannot germinate, or if it germinates, it may produce a weak seedling. The moisture content of the seed may be so high that heating will occur before the seed is dried to a safe moisture content; it may be damaged during drying, or it may be so dry that it suffers impaction damage on handling. The potential danger during the drying process is exposure of very wet seeds to high temperatures or to temperatures in the critical range for a long time.

In the cleaning and the grading or sizing processes the seed may again be subjected to impaction damage, which may be caused by low moisture content in the case of bean and many other crops harvested under arid or semiarid conditions. Radish seed may be damaged severely by impaction if its moisture content is too high during threshing. High moisture makes the seeds difficult to remove from their pods.

When transported to market, whether a few miles or across the ocean, the seed lot is subject to storage under various adverse conditions—on the truck, in the railway car, on the dock while partially or entirely exposed to the elements, or in the hold of ships. A seed lot for use in the following year or years would normally be placed in a conventional storage facility, which may or may not be conducive to maintaining vigor and viability. Because of the high capital investment and cost of maintaining desirable seed storage facilities, many seed lots are stored under unfavorable conditions.

During processing and shipment, grain for industrial uses is frequently moved through suction-type conveyors at great speed and force, impacting with metal at curves and bends, and falling great distances. This results in considerable breakage and other damage. The amount of this damage can usually be calculated and is tolerated, since the grain will be processed for food, feed, oil, or other industrial uses. Grain or other types of seeds to be planted should not be subjected to such treatments. Damage from impaction and improper drying or storage may or may not be obvious on casual observation. Seeds may receive internal fractures from impaction, moisture stress, or heat stress without corresponding damage to the surface.

Decline and Death of Seeds

This publication is concerned primarily with the various factors affecting the life and death of seeds in storage. Decline in vigor and death of seeds can be considered from two aspects: (1) Loss of viability or death of a seed lot, i.e., a small or large quantity of seed, or (2) death of an individual seed. Ordinarily people are interested in the viability of a quantity of seed as this is what they most likely will plant. They want to know whether the germination capacity of the seed lot is sufficiently high to produce a satisfactory stand of plants and, if so, the required rate of planting to produce a stand. The germination percentage of a seed lot is the proportion of individual seeds capable of producing normal plants. For this reason the decline in vigor and eventual death of an individual seed should be considered.

Seeds of most crop species are mature when they attain maximum dry weight. Most seeds are physiologically mature at this point, but there are exceptions. When physiologically mature, the seed possesses its greatest vigor. From this point on, it gradually loses vigor and eventually dies. The rate of decline is conditioned by several factors, some of which man can manipulate. Although he can prolong the deterioration process, he has not been able to stop it.

Since death of a seed is gradual, it is virtually impossible to determine when life ceases. Not all the life processes proceed at the same rate in different seeds of a given seed lot. When seeds are stored under various conditions, these life processes can terminate in individual seeds at different times. For example, certain enzymes remain functional for a time after a seed is dead.

Woodstock *(1973)*[3] expressed biological deterioration of seeds as follows: "In cellular terms there is no sharp line of demarcation between life and death in seeds. Rather, seed death is a gradual and cumulative process as more and more cells die until certain critical parts of the seed become unable to perform their essential function."

Moore *(1955)* summarized the transition from seed maturity to death as follows: "The so-called vigor of life is at its greatest level when the seed first matures. After maturity, however, that life becomes less and less vigorous as the never-ceasing aging process moves onward toward death; but even long before death the seed becomes questionable or worthless for planting purpose, especially under field conditions that are not highly favorable for germination and seedling development.

"This transition of life at its fullest, to death with the ever associated loss of vigor as intermediate steps, may take place slowly or rapidly, depending upon the kinds of seed, the extent and location of mechanical injury, the moisture and temperature conditions to which the seeds are subjected in the field, during harvest, and in storage, and the important factor of time during which the seeds are subjected to these conditions. It is this set of conditions that determine the 'true planting age' of the seed. Seed may thus be surprisingly old, or dead, at a few days or weeks of age, or distinctly young many years after maturity."

Scope of Publication

Publications on seed storage began to appear in the second quarter of the 19th century. At least three or four were in French and German in the 1830's. From about 1840 to 1875 the publications per decade were slightly less than in the 1830's but generally greater in depth. From about 1875 until today the published reports have become so numerous that it is literally impossible for one person to review the literature on seed storage. There seems to have been a great proliferation of papers from about 1890 to 1910, when more and more were published in English. In addition to gaining practical knowledge on seed storage, investigators became interested in determining the longevity of seeds. Some tested the viability of dry seeds taken from old *herbarium* specimens; others set up longtime experiments by burying seeds in the soil in such a manner that they could be recovered for future testing. Owen *(1956)* and Barton *(1961)* have described these experiments.

We have assembled information here that we hope will be useful to those interested in storing seeds of vegetables, flowers, and field crops. Included is information on the principles of seed storage and their

---

[3]The year in italic after an author's name refers to Literature Cited, p. 244.

practical application. Seeds of forest trees, fruit trees, and weeds are not thoroughly discussed. For information on longtime storage of seeds for retention of germ plasm, see James *(1967, 1972)* and Ito *(1972)*.

Many changes occur in seeds during storage. The principal changes considered in this publication are three quality factors—germination capacity, vigor, and yield potential. Since much information exists on chromosome aberrations and mutagenic changes in aged seeds as well as biochemical changes related to enzymes, enzymatic action, and metabolites of these processes, these subjects are treated only briefly.

SEED FACTORS THAT AFFECT STORAGE LIFE

The seed's characteristics, inherent conditions resulting from natural causes, and man's treatment and handling affect the seed's storage life. The storage potential of an individual seed is influenced by the following 10 characteristics and conditions and possibly by others.

Genetic Effects

Variation Among Species

Man learned early that seeds of some plant species would survive a given set of storage conditions longer than others. This information is so common today that it may appear unnecessary to mention it here; however, a few examples are documented.

Essentially all known cases of seeds surviving for a century or more belong to species with hard impermeable seedcoats. Harrington *(1972)* in his list of species with longevity records of 10 or more years included four genera in this group: *Albizia* 147 years, *Cassia* 158 years, *Goodia* 105 years, and *Trifolium* 100 years. All belong to the Leguminosae, a plant family known for producing hard seeds. An exception is a report by Aufhammer and Simon *(1957)* that barley, which had been sealed in glass tubes and placed in the foundation stones of a building in Nuremburg, Germany, showed 12-percent germination after 123 years. Most species known or estimated to have survived for over 500 years are hard seeded; for example, species of *Canna* (Anonymous, *1968*), *Lotus* (Ohga, *1923*), and *Lupinus* (Porsild et al., *1967*).

In contrast to these leguminous seeds, the seed of peanut, also in the legume family, is relatively short lived. Seeds of economic species with a short storage life include lettuce, onion, parsnip, and peanut. Of the cereals, barley and oats usually have the greatest potential for satisfactory storage, rye probably the least, and corn (maize) and wheat are intermediate (Haferkamp et al., *1953*). Haferkamp et al. *(1953)* found that three lots of 32-year-old barley germinated 96, 80, and 72 percent, whereas seeds of four lots of oats stored under the same conditions germinated 84, 66, 55, and 6 percent. MacKay and Tonkin *(1967)* found barley to have about the same storage capacity as wheat and considerably less than that of oats.

Variation Among Cultivars

Inheritance of seed longevity is not limited to species. Several studies show that the principles prevailing at the species level are also effective at the cultivar level, at least for some crop species. Several cultivars of the same species that were compared for seed longevity differed significantly. Shands et al. *(1967)* reported that seeds of the cultivar Oderbrucker barley had greater resistance to germination loss in storage than other cultivars. Toole and Toole *(1954)* found that the cultivar Black Valentine bean had a longer lifespan in storage than 'Brittle Wax.' James et al. *(1967)* found significant differences in the storage of various cultivars of bean, cucumber, peas, sweet corn, and watermelon. Their review of relevant literature indicates that the work of Lindstrom *(1942)*, Sahadevan and Rao *(1947)*, and Haber *(1950)* consistently supported the probability of dominance of longevity. In a study of eight soybean cultivars, Burgess *(1938)* found germination variations between 21 and 99 percent after 4 years compared with 95 to 99 percent after 5 months' storage. In this study the differences among cultivars did not show up until the third and fourth years of storage.

Results of studies on inbred lines of corn by Neal and Davis *(1956)* suggest that some inbred lines maintain viability better than others when stored at low temperatures. Lindstrom *(1942, 1943)* found that some inbred lines of corn germinated 90 percent after 12 years' storage at room conditions in Iowa, whereas all seeds of other inbreds were completely dead. As indicated previously, he also found that increased longevity in corn inbreds appeared to be dominant; however, he also pointed out that inheritance of longevity is not simple. He believed it is possible to introduce genes for longevity into various lines of corn by backcrossing. Weiss and Wentz *(1937)* reported short seed longevity and low vigor of corn to be associated with $luteus_2$ and $luteus_4$ genes found on chromosome 10. Other luteus genes on the same chromosome did not decrease storage life of the seeds. The authors were unable to explain this apparent contradiction.

Generalizations

Not all seed species, cultivars, or individual seeds within a genetic group are destined to survive for the same period of time under a specified set of conditions. A lot or sample of seeds does not die at one time, but the individual seeds making up the lot or sample die over a period of time. In referring to storage life, lifespan, period of viability, or storage potential, we mean the length of time required for a certain percentage of the seeds to die or conversely for a certain percentage to live. Different percentages have been used for different purposes (Roberts, *1972*).

In summary, the seeds of different plant species vary widely in their lifespan under identical or favorable storage conditions. Only a few studies have been made to determine the inheritance of long storage life among cultivars and most of them have pertained to corn (maize). This might be a worthwhile subject for further study. In planning farm or commercial storage the expected storage life of the species should be considered. In experimental studies every effort should be made to use homozygous cultivars, lines, or strains, and studies should be conducted under strictly controlled conditions.

Preharvest Effects

Austin *(in* Roberts, *1972)* reviewed the literature on the effects on viability of the environment during seed development. Included were such factors as temperature, photoperiod, mineral nutrition, rainfall, and soil moisture. However, he did not relate these studies to seed storability. In fact, few, if any, of the studies we have reviewed were designed to test the relative effects of preharvest factors on seed storability.

How Seeds Are Affected

Desirable candidates for storage are mature seeds of normal size and appearance and relatively free of mechanical injuries and storage micro-organisms. They should not have been subjected to extreme temperature and moisture conditions during maturation and harvest. Thus, any preharvest environmental factor that influences these seed qualities also affects the storability of seed lots. About 95 percent of the total dry weight of a seed is storage material, which will be used by the germinating seed and seedling until it can produce its own photosynthetic and nutrient absorbing machinery (Pollock, *1961*). It is readily seen that an immature seed, a seed with an unbalanced chemical composition, or one mechanically damaged, permitting early entry of micro-organisms, would be at a disadvantage in storage.

Barton *(1941)* found that seeds of high initial viability resisted unfavorable storage conditions better than similar seeds of low initial viability. Working with Austrian winter pea, McKee and Musil *(1948)* showed that seed germinating 94 percent did not deteriorate under open storage at ambient temperatures and humidities at Beltsville, Md., and four locations in the Southeastern States as rapidly as similar seed that germinated 75.5 percent at the beginning of the experiment. Brett *(1952a, 1952b)* reported that seeds of grasses, legumes, vegetables, and root crops grown in England with high initial germinations deteriorated less rapidly than similar seeds of low initial germinations. MacKay and

Tonkin *(1967)* checked Brett's work by using large numbers of samples stored under laboratory conditions at Cambridge, England. They confirmed Brett's results for seeds of grasses, root crops, and vegetables but not for legumes. They concluded that it was unlikely that hard seeds in legumes accounted for this behavior.

Bunch *(1958)* adequately summarizes this discussion: "Previous treatment of the seed, in the field, during harvest, or in previous storage may have affected the seed in such a way as to lessen its chances of storing well. In warm and humid climates the seed may deteriorate on the plant before harvesting. Immature seeds and mechanically damaged seed are especially liable to lose viability rapidly in storage."

Effect of Provenance

Very little information is available on the effect of provenance, or location of production, on seed vitality and apparently none relating provenance directly to storability of seeds. MacKay and Tonkin *(1967)* compared the time required for seeds of four forage species grown in different countries to deteriorate to 80-percent germination. The results were as follows: Red clover seed grown in Canada required 4 years compared with 3 years for English- and New Zealand-grown seed; perennial ryegrass seed grown in Ireland and New Zealand required 4 years; American-grown meadow fescue required nearly 7 years compared with 6 years for Danish-grown seed; and Irish-grown crested dogtail seed required 6 years compared with 3 years for New Zealand-grown seed. These data may show significant differences among locations of production; however, the seeds were not stored under controlled conditions and sources of seeds were not predetermined; rather, the data were limited to the seed receipts at the Official Seed Testing Station at Cambridge, England. Thus, no definite conclusion can be drawn at this time as to the effect of provenance on storability of seeds.

Effect of Weather

The most obvious preharvest factor affecting seed viability and storability is weather, especially seasonal changes. Farmers and seedsmen alike know the perils and risks of excessive moisture and freezing temperatures during the later stages of seed maturation and postmaturation stand in the field. Some documentary evidence follows:

Dillman and Toole *(1937)* stored seed of four flax cultivars grown under irrigation in California in 1929 and 1930. The 1930 seed "showed marked weather injury and a low test weight (36.4 to 45.0 pounds per bushel)." The germination of four cultivars after 6 years of storage of the 1930 crop was 1, 4, 0, and 9 percent, whereas the corresponding germination of the 1929 crop of the same cultivars was 94, 86, 87, and 94 percent.

MacKay and Tonkin *(1967)* showed positive correlations between weather conditions during ripening and harvesting of barley, oats, and wheat and the number of years required for the seed to deteriorate to either below 80-percent or below 50-percent germination. The data were obtained from large numbers of samples grown in England over a period of 26 years. Average hours of sunshine and inches of rainfall at Cambridge were used as indices of weather.

Results obtained on seeds of numerous species stored in an unheated building up to 33 years at ambient conditions near Lind, Wash., led Haferkamp et al. *(1953)* to conclude that healthy matured seed stored unthreshed and harvested in dry weather is a prerequisite for long storage life. Riddell and Gries *(1956)* suggested that variations in growth of spring wheat from seed of different ages were related to temperature of maturation rather than to age of seed or storage conditions. Harrington and Thompson *(1952)* found that the region where lettuce seed was grown significantly affected its germination at between 24° and 30° C. Highly significant positive correlations were established between germination percentages and the temperature 10 to 30 days preceding harvest.

Moss et al. *(1972)* found that preharvest rains can cause wheat to germinate in the ear accompanied by increased activity of α-amylase, which digests some of the starch. Fortunately there appears to be some variation among cultivars for activity, suggesting the possibility of selecting or breeding varieties with resistance to this characteristic.

Early sustained freezes in the Corn Belt while seed corn still in the field has a high moisture content can cause serious damage. The degree of damage is affected by the temperature reached, duration of low temperatures, and moisture content of the seed. Effects of the damage may become obvious soon after a hard, sustained freeze. Some seeds in such a lot or entire seed lots exposed to less severe freezes may be damaged only to the extent that the effects are detectable after storage. Because of the seriousness of the problem and its importance, corn subjected to freezing has been investigated to a considerable extent. Some other crop species have been studied also but on a smaller scale. For additional information, see page 30.

Seed Structure and Composition

The presence or absence of glumes (lemma and palea) in grasses is a well-known example of a seed structure that influences lifespan. Haferkamp et al. *(1953)* found that aged seeds of three cultivars of barley and Red Winter Speltz wheat with hulls retained viability better than seeds of the same harvests that had been threshed and stored. Other cultivars of barley and wheat, as well as oats and rye, showed the same trend but to a less extent. Hulls, chaff, or both had an inhibitory effect on the growth of mold, suggesting that the increased lifespan of cereal seeds was due to suppression of mold growth by the glumes during storage. Lakon *(1954)* showed that oat and timothy seeds had a longer lifespan when stored with the glumes intact compared with similarly stored machine-hulled caryopses. Seeds of both species behaved similarly. From Lakon's studies the hulled seeds appeared to be damaged mechanically during removal of the hulls.

Working with seeds from six districts in Sweden, Esbo *(1954)* reported that the viability of hulled timothy seeds declined 16 percent during the first year, whereas seeds with the hulls intact showed no significant loss in viability until the third year. Also, he found that field emergence of unhulled seed was 6 to 14 percent higher than emergence of hulled seed. Average germination of six seed lots after approximately 2 years was 78 percent for unhulled and 40 percent for hulled seed.

Goff *(1890)* found that hulled timothy seeds, on sale or stored in Wisconsin, did not germinate as well the first year nor after storage up to 10 years as did unhulled seeds. The differences were as follows:

Age of seed (years)	*Percent germination*
1	9
5	18
8	24
10	7

The smaller difference for seed stored 10 years may have been because the unhulled seed declined in viability after 8 years of storage. Although Goff used samples from commercial seed lots in his studies, he suggested that the reduced germinability of hulled seeds could not have been caused by thresher injury "as all seed of either sort that showed the least indication of injury were rejected from the trials."

Stevens *(1935)* stored unhulled and hulled timothy seeds under the same conditions for 11 years. During this period the germination of unhulled seed declined from 98 to 52 percent, whereas the hulled seed declined from 97 to 16 percent.

Since the work of Goff and Stevens, several workers have shown that many kinds of seeds suffer internal damage not always apparent without dissection and special tests (Esbo, *1954*). Esbo pointed out that when the loosely attached palea and lemma are detached from the caryopsis, the latter is vulnerable to mechanical injury. He definitely associated the rapid decline in viability of hulled seeds with mechanical injury.

Roberts *(1972)* discussed the protection offered several shapes of seed to mechanical damage during harvesting, handling, and processing. Generally small seeds escape injury, whereas large seeds are more likely to be extensively damaged. Size, arrangement of essential seed structures, and seed composition are contributing factors. Bean and lima bean are examples of seeds highly susceptible to damage. The large cotyledons and location of the embryo axis represent a structure that will tolerate only low level impacts. Although corn is a relatively large, flat seed, its strong pericarp, strong adherence of parts, and location of embryo offer some protection, yet injuries can be expected to the root tip, since it is only moderately protected. According to Roberts *(1972)*, spherical seeds usually give more protection than flat or irregularly shaped seeds. Of course, the spherical shape is mediated by other characteristics, such as relative exposure of vital parts (root and shoot axes), thickness and strength of seedcoat, and brittleness of all seed parts at the time of impact.

Justice *(1950)* referred to the relatively high incidence of abnormal seedlings in onion, as did Clark *(1948)*. Roberts *(1972)* observed that the embryonic root tips of onion extend beyond the seed, a condition conducive to mechanical injury.

Available literature does not relate seed composition or biochemistry to lifespan. Conceivably some relationships may exist, but more definite information is needed. Chemical, physical, and nutritive changes during storage are discussed by Zeleny *(in* Anderson and Alcock, *1954)* and biochemical indices of deteriorating seeds by Abdul-Baki and Anderson *(in* Kozlowski, *1972)*.

Hard Seeds

Much has been written about hard seeds, or seeds that are impermeable to water within the usual germination period, but little information is directly related to seed storage.

Most crop species in the legume or pea family (Leguminosae) produce hard seeds of varying percentages. Harrington *(1916)* indicated that the important legumes cultivated in the United States probably produce varying percentages of impermeable seeds except the peanut. Many of the current crop species can develop hard impermeable seeds when produced and stored under certain environmental conditions. Although hard seeds are usually associated with species of the legume family, they may also be produced by species in other plant families. For example, hard seeds are frequently found in hollyhock and okra and infrequently in cotton in the mallow family (Malvaceae), catnip in the mint family (Labiatae), cranesbill in the geranium family (Geraniaceae), canna in the Cannaceae, *Ipomoea, Convolvulus,* and *Cuscuta* in the morningglory family (Convolvulaceae), and Indian lotus in the Nymphaeaceae.

Possibly impermeable seeds are produced by several species in additional families, especially by the native wild plants. In fact, Passerini

(1933) explored this possibility, but his data fail to distinguish between retarded germination caused by impermeable seeds and physiological dormancy, which may be unrelated to seedcoat effects.

Most of the literature does not state clearly the developmental stage in which the different legume seeds become hard. According to Lowig,[4] fully ripened seeds of white clover contained slightly more hard seeds than less mature seeds. Lute *(1928)* found that the percentage of impermeable seeds of alfalfa increased with maturity. Harrington *(1915)* indicated that 90 percent or more of well-matured seeds of alsike clover, red clover, white clover, and white sweetclover were hard. It appears evident that Harrington referred to the condition of seeds before storage. Martin *(1944)* found that most lots of red clover seed obtained from 13 growers in Iowa contained a considerable number of hard seeds, varying from 0 to 35 percent. Also, Martin *(1945)* reported that sweetclover seed lots coming from the harvester contained, on the average, more than 50 percent of hard seeds. Under certain environmental conditions the hard seed content of some lots increases during storage.

Helgeson *(1932)* showed that practically all slightly immature sweetclover seeds were permeable when harvested but were made impermeable by drying them over calcium chloride for 7 days or without the calcium chloride if kept in dry storage. Impermeability or seed hardness was prevented by storing the seeds in a moist, cold room. Generally a warm, dry atmosphere is conducive to hard seed formation and a cool, moist atmosphere favors a low hard seed population. In many species, hard impermeable seeds have a longer lifespan than permeable seeds. This is an obvious advantage to the survival of the species but causes problems in cultural practices. The current tendency with annual crops is to plant at a rate that will produce the number of plants desired in a given area, as uniform germination is necessary to produce uniformity of crop, permitting "once-over harvesting." Hard impermeable seeds that germinate at some future date have little planting value under these circumstances. On the other hand, storage of seed at intermediate or high relative humidity to minimize the percentage of hard seeds in a lot contributes to more rapid deterioration of germination capacity, vigor, or both.

Varying percentages of hard seed of most species become permeable during overwinter storage in the temperate zone. Softening depends largely on temperature and relative humidity. Harrington *(1916)* conducted extensive studies on the hard seed conditions in small-seeded

---

[4] LOWIG, K. H. E. A STUDY OF METHODS FOR MAINTAINING THE GERMINATION OF SEEDS OVER LONG PERIODS OF TIME—JUNE 1, 1965, TO MAY 31, 1970. Final report on Public Law 480 Project No. E10–MQ–1(a), Grant No. FG–Ge–106. 180 pp. 1970. [Processed.]

legumes. He concluded that nearly all hard seeds of alsike clover, white clover, and sweetclover in dry storage remained impermeable for at least 2 or 3 years. Impermeable red clover seeds gradually became permeable, but 30 to 65 percent of the hard seeds remained impermeable for 4 years or longer. Most alfalfa and hairy vetch seeds became permeable within 2 years.

Martin *(1945)* in Iowa found that up to 24 percent of the hard seeds of sweetclover softened when stored at temperatures around freezing, but softening was nil at 15°–30° C. This confirmed the observations of Stevens and Long *(1926)*. Presumably the seeds were stored at ambient humidities in Iowa. Martin *(1944)* reported that hard seeds of red clover from 13 sources softened to a maximum of 33 percent, depending on source, when stored dry in a laboratory at Ames, Iowa. Two seed lots changed from a relatively low grade, based on germination percentages, to a fair grade during a short period of storage. According to Lute *(1928)*, 50 percent of the hard seeds of alfalfa became permeable within 3½ years when stored in Colorado.

Researchers have consistently reported great differences in the percentages of hard seeds among species and among seed lots of the same species (Whitcomb, *1931*; Esdorn and Stuetz, *1932*). Esdorn and Stuetz *(1932)* reported that hard seeds of lupines, alfalfa, and red clover softened rapidly under the suitable temperature-humidity conditions, whereas softening in hard seeds of alsike clover and white clover proceeded very slowly.

Much has been written about the structure of leguminous seeds in relation to water absorption. A few references are cited here. *Seedcoat* or *testa* (Rees, *1911*; Watson, *1948*), *cuticle* (White, *1909*), *hilum* (Hyde, *1954*), *strophiole* (Martin and Watt, *1944*), and a general paper by Coe and Martin *(1920)*. The anatomy of some other crop species that produce hard seeds is discussed by Simpson et al. *(1940)* for cotton, Reeves *(1936)* for cotton and other malvaceous plants, and Gaertner *(1950)* for *Cuscuta* (Convolvulaceae).

The following genera and number of species (in parentheses) of field crops, vegetables, herbs, and flowers can be expected to produce hard seeds. They are included in the "Rules for Testing Seeds" (Association of Official Seed Analysts, *1970*).

Compositae (aster)
Centaurea—basketflower (1)

Convolvulaceae (morningglory)
Dichondra—dichondra (1)
Ipomoea—morningglory (3)

Geraniaceae (geranium)
Geranium—geranium (1)

Labiatae (mint)
Nepeta—catnip (1)

Leguminosae (legume)
Alysicarpus—alyceclover (1)
Coronilla—crownvetch (1)
Crotalaria—crotalaria (5)
Cyamopsis—guar (1)
Desmodium—beggarweed (1)
Dolichos—hyacinth-bean (1)
Glycine—soybean (1)
Indigofera—hairy indigo (1)
Lathyrus—roughpea, sweetpea (3)

Leguminosae (legume)—Con.
Lens—lentil (1)
Lespedeza—lespedeza (4)
Lotus—trefoil (2)
Lupinus—lupine (6)
Medicago—alfalfa, black medic, burclover, buttonclover (5)
Melilotus—sweetclover (3)
Mimosa—sensitive plant (1)
Onobrychis—sainfoin (1)
Phaseolus—bean, lima bean (5)
Pisum—pea (1)
Pueraria—kudzu (1)
Sesbania—sesbania (1)
Stizolobium (now *Mucuna*)—velvetbean (1)
Trifolium—clover (14)
Vicia—vetch (7)
Vigna—asparagusbean, cowpea (2)

Lilaceae (lily)
Asparagus—asparagus (2)

Malvaceae (mallow)
Althaea—hollyhock (1)
Gossypium—cotton (1+)
Hibiscus—hibiscus, okra (2)

Some snap bean cultivars produce hard seeds, although most do not. Hard seeds in snap bean are undesirable. Among cultivars containing hard seeds are Top Notch and Golden Wax (Barton, *1966a*), Decatur and Blue Lake (Nutile and Nutile, *1947*), White-Seeded Kentucky Wonder (Harrington, *1949*), and Green Savage (Joubert, *1954*).

Lowig[5] listed 41 cultivars of bean, mostly of German origin, with hard seeds varying between 0 and 97 percent. Thirteen of these showed hard seeds occurring at 50 percent or above, and 21 cultivars showed hard seeds exceeding 25 percent. These figures must be used with caution since Lowig determined hard seed percentages after storage over calcium chloride.

Researchers in Puerto Rico (Anonymous, *1942a*) found that drying beans with calcium chloride was unsatisfactory because of the hard seeds produced. Nutile and Nutile *(1947)*, Harrington *(1949)*, and Barton *(1966a)* demonstrated the effect of low relative humidity on hard seed formation in bean. For example, Harrington stored White-Seeded Kentucky Wonder beans with an initial hard seed content of 33.5-percent and 8.3-percent moisture content. After 60 days of storage at 10-percent relative humidity, 74.4 percent of the seeds were hard, whereas samples stored at 65-percent relative humidity for the same time contained no hard seeds.

It is obvious that seed of certain snap bean cultivars can change readily from permeable to impermeable depending on the relative humidity of the storage area. Such is not so with seed of other species. Seed of several small-seeded legume species became hard when stored at a low relative humidity and remained hard when transferred to a high relative humidity because the valve-type closure to the hilum permitted moisture to escape from but not enter the seed. When impermeable seeds of most species became permeable, they remained so regardless of the relative humidity where they were subsequently stored.

In some trade practices, hard seeds are regarded as viable. Several

[5] See footnote 4, p. 14.

scientific papers attest to the fact that not all hard seeds are viable, for example, Goss *(1926)* and Toole *(1939)* for hairy vetch. Crosier and Patrick *(1952)* found the viability of hard and nonhard hairy vetch seeds to be similar during the first 5 years of storage, but after 17 years both were among the dead ones. Also, dead hard seeds have been reported in okra.

Nutile and Nutile *(1947)* and Harrington *(1949)* made suggestions for storing bean seed that may have become hard with age. Nutile and Nutile recommended that such lots not be stored in artificially heated buildings. Harrington recommended that storing such seeds at about 21° C and a relative humidity of 50 percent or possibly a little higher should prevent hard seed formation except possibly for very low percentages in the most susceptible cultivars.

Seed Maturity

Relationship of Maturity to Storability

Several scientists have considered seed maturity as being that stage when maximum dry weight has been attained (Harrington, *1972*; Roberts, *1972*). Since many crop species flower, produce, and mature seeds over a period of several days or even weeks, it is important to know at what stage of maturity the seeds should be harvested. Much of the research relating to maturity has been oriented toward production and harvesting goals without continuing into storage studies. Such studies have marginal relevance to the relationship of seed maturity and lifespan in storage. Since healthy, mature, plump seeds generally store better than immature seeds, any studies that show the relationship between seeds of different maturity levels and germination, vigor, field emergence, or yield relate indirectly to storage and lifespan. Many such studies have been discussed by Harrington *(1972)* and Austin (*in* Roberts, *1972*).

Grass Seeds

Several studies on the relationship between seed maturity and storability or longevity have been concerned with grasses. Both the flowering and seed maturity characteristics of the grass plant may extend over several days depending on crop species. This results in harvesting seeds of varying stages of maturity from recently fertilized ovules to mature, plump seeds. Many of the light grass seeds are removed in the cleaning processes; however, many are not, especially those in the premilk and milk stages.

McAlister *(1943)* collected seeds of three species of *Agropyron*, three species of *Bromus*, *Elymus glaucus*, and *Stipa viridula* in the premilk, milk, dough, and mature stages and stored them at between 15° and 23°

C at moisture contents of 7–9 percent. Evaluation tests made after 4 and 5 months showed that the seeds harvested in the premilk and milk stages were generally inferior as to both viability and longevity to seeds harvested in the dough and mature stages; the only exceptions were that viability of seeds of *Bromus marginatus* and *B. polyanthus* harvested in the premilk and milk stages was equal to that of seeds harvested in the dough and mature stages.

Similar results and conclusions have been reported by numerous workers, including Hermann and Hermann *(1939)* for *Agropyron cristatum*, Griffith and Harrison *(1954)* for *Phalaris arundinacea*, Esbo *(1959)* for *Phleum pratense*, Bass *(1965)* for *Poa pratensis*, and Jorgensen *(1969)* for *Festuca pratensis*.

Other Field Crop Seeds

Only limited work has been done on seed maturity and storage of large seeds. However, studies have shown in many instances that immature or partially filled seeds are inferior to mature seeds in viability and vigor. These include Dimmock *(1947)* for corn, Blackstone et al. *(1954)* for peanut, Inoue and Suzuki *(1962)* for bean, and Turner and Ferguson *(1972)* for cotton. Justice and Whitehead *(1942)* found immature velvetbean seed to be inferior in viability to mature seeds. The germination for nine sublots that could be compared were mature seeds 73 percent and immature seeds 10 percent. Storage period varied with sublots from 3 to 7 months depending on the date of harvest.

Vegetable Seeds

Two Japanese workers reported some relevant information on the effect of maturity on vegetable seed storage. Eguchi and Yamada *(1958)* stored seeds of 11 kinds of vegetables (cabbage, carrot, Chinese cabbage, cucumber, edible burdock, eggplant, Japanese radish, pumpkin, tomato, watermelon, and Welsh onion), which they had harvested at 3- to 7-day intervals giving 4 to 6 stages of maturity for 2 to 3 years. Their English summary indicates that eight kinds showed marked losses in viability of immature seeds compared with mature seeds. However, Chinese cabbage, pumpkin, and tomato showed little or no difference in longevity with different maturities.

Pollock *(1961)* pointed out that maturity studies of carrot are complicated by the fact that at least three orders of flowers, or umbels, are produced and each order will flower and produce seeds over a period of time. Harvesting all umbels at the same time will result in both mature and immature seeds. He reported that the level of maturity affects both germination and vigor.

Austin *(in* Roberts, *1972)* harvested carrot seeds in England at four dates from September 5 through October 5 and classified each harvest into four size groups. As the harvest season progressed, the weight of seeds declined slightly, but the germination increased by 5 percent, which probably was not significant. As the size of seed increased, the mean seed weight increased correspondingly. The germination increased with increased seed size for the first three weight classes (1.00 to 1.75 mm per seed) and for each harvest date, but it decreased by 4 percent for the largest and heaviest seeds (1.75 to 2.00 mm per seed). These studies suggest that immature carrot seeds do not germinate as well as mature seeds. Based on studies with seeds of other crop species, the longevity of immature seeds would be apt to be less than that of mature seeds when both are stored under identical or similar conditions.

Germination response of seeds removed from mature and immature squash fruits differed presumably because of seed maturity. Young *(1949)* and Holmes *(1953)* collected mature and immature fruits of butternut squash and stored them for periods up to about 7 months. At the beginning of his experiment, Young found seeds from mature and immature fruits to weigh 7.78 and 5.32 gm per 100 seeds, with average germination of 90.8 and 19.0 percent, respectively. Four months later, both mature and immature seed had gained weight by 0.34 and 0.91 gm and germination had increased to 98.4 and 67.2 percent, respectively. At the beginning of the experiment, seeds from only 7 out of 25 immature fruits germinated, whereas all immature fruits produced germinable seeds at the conclusion of the experiment. This indicates that the seeds of some immature fruits either were too immature to germinate when harvested or were dormant.

After 7 months of storage of butternut squash, Holmes *(1953)* found the weights per 100 seeds to be 10.9 gm for mature and 8.9 gm for immature seeds. Germination at this time was 97 percent for mature and 77 for immature seeds. Holmes *(1953)* also found that the ratio of fruit weight of typical butternut squashes to seed weight was rather consistent.

Seed Size

No references have been located relating seed size or weight to lifespan through storage experiment. On the other hand, numerous studies have shown the superiority of heavy, mature seeds over light, immature seeds in germination, vigor, and yield tests. Relatively few exceptions have been noted. Comprehensive reviews are provided by Black *(1959)* and Austin *(in* Roberts, *1972)*. Austin *(1972)* discussed seed maturity and seed size jointly, possibly because seed size is determined or mediated to some extent by maturity. His treatment lends credence to the view that the same or similar vagaries besetting immature seeds in storage also affect small seeds, at least to some extent.

Seed Dormancy

Relationship of Dormancy to Storage

Some or all seeds of several crop species are dormant at harvest. A major problem facing seed technologists when testing stored seed for germination is how to overcome dormancy. Dormancy of freshly harvested seeds may be found in practically all groups or classes of plants, whether crops or native plants, including grasses, cereals, clovers and other small-seeded legumes, large-seeded legumes including peanut, cucurbits, vegetables, flowers, trees, and weeds. In some seeds, dormancy is caused by (1) seed structures, as seedcoats, bracts, glumes, pericarp, and membranes, which limit the exchange of water and gases, (2) physiology of the embryo, (3) germination inhibitors or other blocks, or (4) a combination of these factors.

Storage can affect dormancy in many instances. In most crop species, dormancy is dissipated within a few to several months when the seed is stored at ambient temperatures and relative humidities or under controlled atmospheres provided the temperature is held above freezing. Seed physiologists know that the best method of maintaining dormancy in seeds is to store them at subfreezing temperatures.

Dissipation of Dormancy

Owen *(1956)* and Koller *(1972)* reviewed pertinent literature on the existence of seed dormancy and its dissipation with time. Only a few relevant studies are mentioned here. Larson et al. *(1936)* found that dormancy in immature seeds of barley, oats, rye, and wheat persisted longer than in mature seeds when stored at 0° to −35° C, but that dormancy varied greatly among cultivars of a given crop species. The duration of dormancy could be increased by lowering the temperature. In a study of seeds of different cultivars of barley, oats, and sorghum, Brown et al. *(1948)* found that incipient dormancy was overcome in most cultivars within 1 to 6 months when stored at 40°. At 2.2°, dormancy in barley and oats persisted for 3 years.

Dore *(1955)* and Roberts *(1972)* discussed dormancy in rice. In Dore's *(1955)* studies of 21 cultivars, dormancy was lost between the 7th and 11th weeks after harvest except for the cultivar Penifun, which was dormant for only 5 weeks. Roberts *(1972)* found the dormancy of rice cultivars to vary between practically none in the cultivar Tai Chu 65 to over 100 days (100 days for 50 percent of the seeds to break dormancy) in Masalaci. Although dormancy among rice cultivars differed markedly, the pattern of viability loss for all cultivars was the same.

The situation with grasses is similar to that for cereals. In many native American grasses used for range planting, seed dormancy is much more intense than in cereals. Winchester *(1954)* found that

Panicum maximum var. Trechoglume seed increased in germination from 11 percent when fresh to 46 percent after storage for 25 months and seed of *Cenchrus ciliaris* (buffelgrass) from 3 to 79 percent within 13 months.

Dormancy in Florida runner peanut seed was overcome by Hull *(1937)* when he stored it at 20°–25° and 40° C, whereas dormancy in some samples persisted for 2 years at low temperature storage conditions. Recently harvested cottonseed can be dormant. Christidis *(1955)* reported that some cottonseed samples grown in Greece remained dormant up to 5 months. Justice *(1956)* found that the only effective method of breaking dormancy in seeds of *Cyperus rotundus* was to place them on a moist substratum at 40° for 3 to 6 weeks.

Dormancy and Lifespan

In considering the relationship of seed dormancy to storage, a second area of interest is whether dormancy increases the lifespan of seeds. Roberts *(1972)* reviewed pertinent literature on the subject and correctly concluded that available evidence is not sufficient to establish even a casual relationship. He detailed and analyzed some of his own experimental work with rice and concluded that his work did not support the existence of a relationship between dormancy and longevity. This does not exclude the possibility that the lifespan of dormant seeds of native or wild plants buried in the soil under natural conditions may be longer than the lifespan of nondormant seeds.

Toole and Toole *(1953)* found that imbibed dormant seeds of lettuce did not decay in 105 days, whereas companion samples placed in beakers with a relative humidity of 85 to 90 percent were dead; the seeds of one cultivar remained alive for 42 days. This suggested to the authors that dormancy of the imbibed seeds either held in check or suppressed the life processes that lead to seed deterioration.

Existing evidence is not sufficient to demonstrate a positive relationship between seed dormancy and lifespan. However, dormant seeds in storage undergo changes, some of which lead to breaking of dormancy and others to inducing dormancy. Some of these changes are affected by storage conditions.

Moisture Content

The moisture content of seeds during storage is no doubt the most influential factor affecting their longevity. It is important to harvest mature, relatively dry seeds or to reduce the moisture content of high-moisture seeds soon after harvest. Bass *(1953)* found that the loss of viability of freshly harvested Kentucky bluegrass seed was correlated with the moisture content of the seed and length of time held at a given temperature. Seed with 54-percent moisture held at 30° C for 45 hours lost 20-percent germination, but seed with 44-percent moisture withstood 45° for 36 hours with no apparent loss of viability. Seed with 22- and 11-percent moisture content showed essentially no loss in viability at 50° for over 45 hours.

McNeal and York *(1964)* studied the effects of drying sorghum seed on viability. They concluded that sorghum to be used for seed should be harvested when the moisture content is about 20 percent or less and dried promptly to about 11-percent moisture content. The drying temperature should not exceed 43.3° C for very moist seed and not over 54.4° for seed with a relatively low moisture content.

Haferkamp et al. *(1953)* reported reduced germination of 27-year-old wheat compared with slightly older wheat. They explained this on the basis that the June rainfall in 1923 (year of harvest of 23-year-old wheat) was over five times the average. The authors believed that high moisture content of the seeds during development greatly reduced seed viability and longevity. Regarding moisture during seed development, McIlrath et al. *(1963)* made an interesting observation on developing tomato seeds. Although the moisture content of the placenta and pericarp of tomato fruits 15 days old or older never dropped below 94 percent, the seeds became dehydrated to approximately 50-percent moisture content as the organs developed and matured.

Another way in which initial seed moisture content affects storage life of seeds is through its effect on mediating damage by threshing and processing machinery and handling. Although it is very important to reduce seed moisture content to a safe level for storage, it is also important to be aware of possible adverse effects of low moisture content. Very dry seeds are susceptible to mechanical breakage and related injuries. Such damage may result in physical breakage or fracturing of essential seed parts, make the seed vulnerable to fungal attack, and reduce storage potential. For more information, see the section on "Effects of Storage Environment on Seed Longevity."

Mechanical Damage

Awareness of Seed Damage

As harvesting and threshing machinery and the combine came into general usage, damage to seeds and grain increased accordingly. Some of the early work on seed injury related to the mutilation of corn (Brown, *1922*), seedcoat injury of barley and wheat (Hurd, *1921*), broken pericarp of corn (Myers, *1924*), and mechanical injury to soybeans (Oathout, *1928*). Research on seed injury was expanded greatly in the 1930's, with the advent and use of the combine thresher. Some seed species or groups have received greater attention than others depending on their vulnerability to damage by machinery and handling.

Two important characteristics that affect the degree of damage are seed structure and resistance of the seed to removal from the pod, as in legumes and crucifers, or from the mother plant, as in the grasses. Problems arising from these characteristics are affected by moisture content of seeds and pods, degree of maturity, and possibly other factors.

Beattie and Boswell *(1939)* and Moore *(in* Roberts, *1972)* showed that damaged seeds do not store as well as intact seeds and that fungi enter the seed through cracks in the seedcoat. In discussing the storage of seeds of the clovers and medics, Brett *(1952a)* stated, "From the results of experiments at Cambridge [England], and elsewhere, it is clear that scarified seed loses its ability to germinate at a significantly greater rate than does unscarified seed when stored under the same conditions."

Natural deterioration in spinach seed has demonstrated that the percentage of abnormal seedlings increases with time and the percentage of normal seedlings decreases. Thus, seedling abnormalities result not only from mechanical damage but also from natural aging. According to Moore *(in* Roberts, *1972)*, small and hidden injuries in seeds, including bruises, may not cause immediate loss in vitality, but they can become increasingly critical with aging of the seed. Moore *(1972)* pointed out that injuries to vital organs, i.e., various parts of the embryo, become more serious with age than injuries to nonembryonic tissues.

Vulnerability of Some Seeds to Damage

The amount of fiber in the bean pod, which varies with cultivars, affects the degree of mechanical damage to the seeds through its effect on ease of threshing. Apparently this problem was first studied by Harter *(1930)* of California, then continued and expanded by his colleagues Borthwick *(1932)* and Bainer and Borthwick *(1934)*. A comprehensive review of the subject by the Asgrow Seed Co. (Associated Seed Growers, Inc., *1949*) also included results of its own extensive research studies. This type of research has been continued since the 1950's, especially at the U.S. Department of Agriculture, Beltsville, Md. (Toole, *1950*; Toole et al., *1951*; Toole and Toole, *1960*) and at Michigan State University (Perry and Hall, *1960*; Dorrell;[6] Dorrell and Adams, *1969*). These studies showed that threshing or combining produces breaks, cracks, bruises, and abrasions in seeds, which in turn result in abnormal seedlings of questionable planting value. These injuries were reproduced by controlled mechanical shock to the seeds and were greater with reduced seed moisture.

---

[6]DORRELL, D. G. SEEDCOAT DAMAGE IN NAVY BEANS (PHASEOLUS VULGARIS), INDUCED BY MECHANICAL ABUSE. 1968. [Unpublished Ph.D. thesis. Copy on file Dept. of Botany, Mich. State Univ., East Lansing.]

Toole *(1950)* attempted to compare the germinability of slightly injured seeds with seeds producing apparently uninjured seedlings of bean stored for 3, 6, and 9 months. Owing to unpredicted changes during storage, such as (1) decrease in apparently uninjured seedlings, (2) corresponding increase in slightly injured seedlings, and (3) increase in abnormal seedlings, no firm conclusions could be drawn. However, the author suspected that under unfavorable conditions slight mechanical injuries could predispose the seed to increased deterioration compared with uninjured seed, even though the preliminary study did not confirm this suspicion.

Response of Seeds to Storage Conditions

There is reliable evidence that damaged seeds of small-seeded legumes do not survive as long as nondamaged seeds. Battle *(1948)* found that all alfalfa seeds that had been scarified with sandpaper and stored for 14 years were dead, whereas unscarified seeds showed 23-percent germination. Similar results were reported by Graber *(1922)*, Stevens *(1935)*, and Brett *(1952a)*. In storage experiments, Oathout *(1928)* and Mamicpic and Caldwell *(1963)* demonstrated that damaged soybean seeds lose their viability earlier than nondamaged seeds. Blackstone et al. *(1954)* reported similar results for peanut.

Information on the retention of viability of injured cereals in storage is extremely limited. In a 17-year experiment Russell *(1958)* compared lots of wheat seed having cracked seedcoats with lots of sound seed. At the beginning of the experiment the damaged seed germinated 4 percent less than the sound seed. This difference gradually increased to 12 percent, where it remained for 9 years of storage. The germination of the two groups differed slightly during the last 4 years of storage. Possibly the sound seed declined rapidly in viability after 10 to 12 years. Hurd *(1921)*, Kuperman *(1950)*, and Wortman and Rinke *(1951)* showed that broken cereal seedcoats provide easy access for microflora to enter the seed. Broken or cracked seedcoats also enhance embryo damage by chemical treatments, including chemicals used as disinfectants (Bonvicini, *1951*). Both fungi and chemical damage reduce the keeping quality of stored seeds.

Possible Involvement of Genetics

Efforts have been made to determine whether there are genetic differences among bean cultivars in their tolerance to mechanical injury. Atkin *(1958)* studied 3 crops of 18 cultivars and concluded that important fairly constant differences existed. Generally the colored-seeded cultivars had greater resistance to mechanical injury than white-seeded cultivars. Nevertheless one white-seeded cultivar (Streamliner) showed some resistance to mechanical injury. This led the author to believe that a resistant white-seeded cultivar could be developed. On the other hand, Dorrell[7] found that tolerance to seed injury of navy bean is a complex trait controlled by numerous genes. Population averages showed no evidence of dominance in the F_1 and F_2 generations and genetic variance appeared to be primarily additive. Thus, from this study there is little encouragement that cultivars highly resistant to mechanical injury can be developed through a breeding program.

It is obvious from available information that mechanical injury to seeds not only reduces production of normal seedlings but also decreases the storage potential of damaged seeds that apparently would have produced normal seedlings prior to storage. Many studies have been made on seed injury resulting from mechanical harvesting, threshing, combining, and handling, but relatively few of these studies have been concerned with storage.

Vigor

The vigor of seeds at the time of storage is an important factor that affects their storage life. Vigor and viability cannot always be differentiated in storage experiments, especially in seed lots that are rapidly deteriorating. Irrespective of this problem, several workers have demonstrated that seed lots that were rapidly declining contained some seeds low in vigor and other seeds that still might be vigorous. This progressive weakening with age continued until all the seeds became nonviable.

Obviously new, vigorous seed lots possess a greater storage potential than older lots that may be approaching rapid deterioration. Generalized curves of the decline in vigor and viability of a seed lot with time are shown in figure 2. The viability curve gradually decreases, followed by a sharp decline and finally a gradual loss. The loss in seed vigor essentially parallels that of viability but at lower levels. The rate at which seeds will decline in vigor or viability depends on several factors, including genetic constitution of the species or cultivar, condition of the seed, storage conditions, uniformity of seed lot, and storage molds if the relative humidity permits their growth. Among the research workers who have shown that vigorous seeds reach a point when deterioration proceeds more rapidly than during the early period of storage are Patrick *(1936)* for Chewings fescue seed, Christidis *(1954)* for cottonseed, and Burns *(1957)* for blue lupine seed.

The decline in vigor and viability of seeds is sometimes illustrated by a sigmoid survival curve. The survival curve for dry seeds stored under favorable environmental conditions can be dissected into three distinct parts. The first represents the period when the seed is vigorous and decline in the life functions has proceeded slowly. Eventually this stage

---

[7]See footnote 6, p. 23.

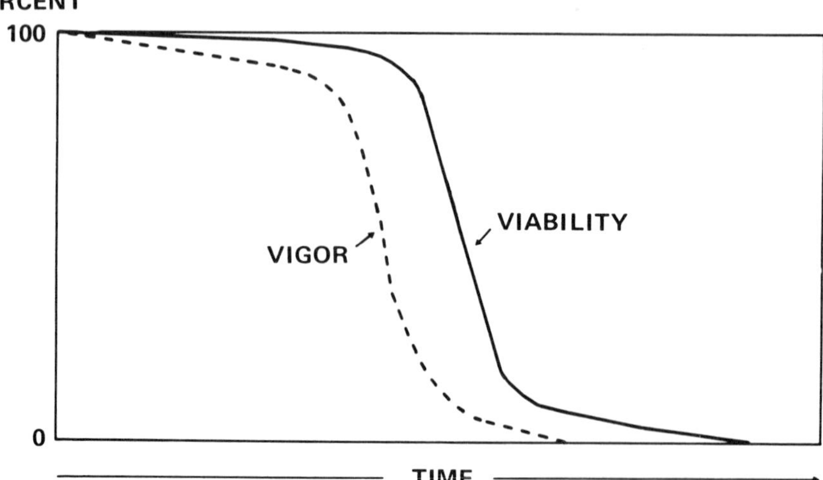

FIGURE 2.—Decline of viability and vigor with time of a typical seed lot. (Courtesy of J. F. Harrington.)

ends at a survival level of 90–75 percent, and deterioration then proceeds very rapidly. After deterioration has proceeded to a survival level of 25 to 10 percent, it slows again and continues slowly until all seeds are dead. The curves for vigor and viability are very similar, except that loss of vigor precedes loss of viability.

EFFECTS OF STORAGE ENVIRONMENT ON SEED LONGEVITY

Temperature

Temperatures Above Freezing

Storage temperature and seed moisture content are the most important factors affecting seed longevity, with seed moisture content usually more influential than temperature. Owing to the intricate relationship between storage temperature and seed moisture content, neither one can be discussed separately in its entirety. Nevertheless temperature and seed moisture are discussed here separately to a limited extent to indicate the relative importance of each, whereas the interaction of the two is discussed more fully under "Interrelationship of Temperature, Seed Moisture Content, and Storage Life."

Within limits the storage life of seeds of vegetables, flowers, and field crops decreases as the temperature increases. Exceptions include certain short-lived seeds. Groves *(1917)* described the relationship between temperature and lifespan of wheat seeds by a formula. Roberts *(1972)* presented formulas that he believed described the relationship of temperature and moisture content to the period of seed viability of certain crop species. He developed nomographs using the formulas for barley, broadbean, rice, and wheat. The accuracy, limitations, and usefulness of these formulas and nomographs are still to be determined.

Harrington *(1960d)* proposed his so-called Thumb Rules, which related seed moisture and temperature to the seed's lifespan. His *(1972)* rules stated that the life of the seed is halved (1) for each 5° C increase in seed storage temperature and (2) for each 1-percent increase in seed moisture content. Harrington cautioned that, based on current knowledge, the first rule should not be used for storage temperatures below 0° or above 50°. He also pointed out that the two rules apply independently, and the effects of temperature and moisture are additive. Since this procedure gives only rough estimates of storage potential under stipulated conditions, it must be used with caution.

San Pedro *(1936)* stored seeds of vegetable species in containers with calcium chloride at four temperatures and determined the percentages of germination after 7.1, 9.2, 11.3, and 19.5 months. His results for the 11-month storage period (table 1) show that as temperature was increased, even in a dry atmosphere, germination decreased. His averages are mediated by the stage of deterioration of each crop species during the storage period. For example, cabbage declined in germination by 19 percent between storage at 0° and 28° C, whereas pechay and sitao declined only 2 and 3 percent, respectively, under the same storage conditions. The seeds of radish were dead at all storage temperatures after 14.5 months.

Barton *(1941)* stored seeds of six vegetable species at different temperatures and relative humidities for 372 days. At 35-percent rela-

TABLE 1.—*Effect of various temperatures on germination of several crop seeds after storage with calcium chloride for 11 months*[1]

Crop	Germination when stored	Germination after storage at indicated temperature (°C)			
		0	10–13	20–22	27.5–28
	Percent	*Percent*	*Percent*	*Percent*	*Percent*
Cabbage	90	72	71	63	53
Carrot	65	57	51	45	42
Lettuce	63	61	51	55	50
Parsley	56	32	32	31	28
Pechay	99	91	89	86	89
Radish	85	23	24	27	9
Sitao	93	83	77	81	80
Average	79	60	56	55	50

[1] Adapted from San Pedro *(1936)*.

tive humidity the average germination for lettuce and onion, two species with short storage lives, was 61, 44, 31, and 9 percent, respectively, at 5°, 10°, 20°, and 30° C, whereas comparative germination results for tomato and flax were 96, 92, 88, and 85 percent.

Barton *(1966b)* also stored pyrethrum seeds at various temperatures in sealed and open containers for up to 15 years. Table 2 shows that temperature markedly affects germination. It also shows that the effect of temperature is influenced by seed moisture content. In sealed storage, seed moisture content remains constant, but in open storage it changes with changes in the relative humidity of the storage area. Therefore adequately dried seeds in sealed containers usually live longer at a given temperature than similar seeds in open containers.

The effect of temperature on the storage of soybean seed was reported by Toole and Toole *(1946)*. They stored seeds of the cultivars Mammoth Yellow and Otootan at five temperatures and three moisture contents for 10 years. The months required to lose germination at each temperature-moisture combination are shown in table 3. Not all seeds were dead at the favorable storage conditions. In fact, there was no significant loss in germination at the end of the 10-year storage period at five storage conditions for Mammoth Yellow and at five conditions for Otootan. The data for 18.1-percent moisture (Mammoth Yellow) and 17.9 percent (Otootan) show the reduction in seed germination or lifespan as temperature is increased, or alternatively the increased germination or lifespan as temperature is decreased. The range in lifespan is from 3 months at 30° C to 72 months at 2° for Mammoth Yellow and from 1 month at 30° to 72 months at 2° for Otootan. Similar data could be compiled for other levels of germination, for example 50 or 80 percent.

TABLE 2.—*Effect of different temperatures on germination of pyrethrum seeds in sealed and open storage after various periods*[1]

Temperature (°C)	Germination in sealed storage after indicated years[2]			Germination in open storage after indicated years[2]		
	3	13	15	3	13	15
	Percent	*Percent*	*Percent*	*Percent*	*Percent*	*Percent*
5	61	53	47	44	44	29
10	53	37	28	38	0	---
20	46	10	6	47	1	---
30	1	---	---	0	---	---

[1]Data from Barton *(1966b)*. Original germination of seed lot was 57 percent.
[2]Dashes indicate no samples were taken.

TABLE 3.—*Effect of various temperatures on duration of seed germination of 2 soybean cultivars at 3 seed moisture contents*[1]

Cultivar	Moisture content when stored	Time required to reach 0-percent germination at indicated temperature (°C)[2]				
		30	20	10	2	−10
	Percent	*Months*	*Months*	*Months*	*Months*	*Months*
Mammoth Yellow	18.1	3	9	24	72	120*
	13.9	5	24	120	120+	120+
	9.4	24	120	120+	120+	120+
Ootootan	17.9	1	5	24	72	120**
	13.4	5	24	96	120+	120+
	8.1	48	120‡	120+	120+	120+

[1] Adapted from Toole and Toole *(1946).*

[2] Percent germination after 10 years' storage: + = above 90, ‡ = 68, ** = 44, * = 17.

Small field plantings were made of 10-year-old seeds that had shown no loss in germination and of seeds grown the previous year. The authors found no apparent difference in vigor of growth for either cultivar when the known genetic variation in Mammoth Yellow was considered. Many seedling abnormalities were observed in both Mammoth Yellow and Otootan cultivars grown from seeds artificially dried to 5.4- and 5.2-percent moisture content, respectively. These abnormalities were in seedlings produced from seed stored for only 3 months, especially from those seeds held at the lower temperatures. Few abnormal seedlings were produced from seed stored at 30° C, but they were increasingly numerous at successively lower storage temperatures. The abnormalities of Mammoth Yellow were much more frequent and more pronounced than those of Otootan. Tests made on seeds stored for longer periods showed approximately the same proportion of abnormalities.

Some research has been done to determine the effects of high temperatures on seeds. In general, seed viability and vigor are reduced as temperature is increased, as time of exposure to high temperatures is increased, and as moisture content of the seed is increased. Damage is diminished at a given temperature as the seed moisture content is decreased. Thus, Waggoner *(1917)* found the following relationship in radish seeds:

Moisture content (percent)	*Nonlethal temperature (°C)*	*Lethal temperature (°C)*
45	Below 50	55–60
4	75	95–100
.4	100	123–125

Harrington and Crocker *(1918)*, Smith and Gane *(1938)*, and Evans *(1957b)* showed that low moisture content Kentucky bluegrass, Chewings fescue, and perennial ryegrass seeds can be heated to 100° C or higher without immediate loss of viability. However, stress tests of heated seeds showed that vigor decreased before the effects on seed viability were manifested. White *(1909)* significantly reduced the percent germination of barley, oats, and wheat by exposing air-dry seeds to dry heat at 99°–100° for one-half hour. Greater decreases were obtained by heating for 4½ hours. Seeds of these three species plus corn (maize) and rye failed to germinate when the temperature was increased to 122° or higher for 1 hour. The consideration of temperature, time of temperature exposure, and moisture content of the seeds is essential when drying seeds for storage. For further information, see the section on "How Seeds Are Dried."

Temperatures Below Freezing

As background for their experimental results from freezing seeds at −183°C, Brown and Escombe *(1897–98)* reviewed at least five papers on low temperature treatments. They referred to papers by C. de Candolle and R. Pictet published as early as 1879 and 1885. They froze seeds at −80° for 2–6 hours, and in 1884 they exposed seeds to −100° for 4 days without adversely affecting germination. Brown and Escombe *(1897–98)* exposed seeds of 12 species with moisture contents of 10–12 percent to −183° for 110 consecutive hours without damage.

White *(1909)* observed severe injury to seeds of apple, hemp, lobelia, parsley, and parsnip after exposure to −200° C for 1½ days, slight injury to seeds of bean, mustard, pea, radish, sunflower, and turnip, and no injury to seeds of cress and *Ricinus camogdensis.* The moisture content was not given. She subjected seeds of barley, corn (maize), oats, rye, and wheat to −200° for 2–3 days without any decrease in germination.

Lipman and Lewis *(1934)* subjected seeds of 19 species to −250° C for 60 days and evaluated the seeds by greenhouse tests. They found no significant difference between the frozen seeds and unfrozen controls. Lipman *(1936)* exposed seeds of barley, corn, sweetclover, and vetch to −273.1° just shy of absolute zero, with no impairment of viability, whereas Becquerel *(1953)* dried seeds of alfalfa, clover, petunia, and tobacco and then held them at −272.9° for 2 hours without impairment of viability.

The superiority of subfreezing temperatures for seed storage compared with higher temperatures has been well established for many kinds of seeds. Documented experimental evidence dates back to 1928 (Dorph-Petersen, *1928*) and possibly earlier. Some of the literature on seed storage at subfreezing temperatures is summarized in table 4.

TABLE 4.—*Sampling of literature on seed storage at subfreezing temperatures*

Literature source	Seeds	Storage		Results (°C)
		Below 0° C	Above 0° C	
Dorph-Petersen *(1928)*	*Tussilago farfara*	−15°	4 temps. (0°–20°)	Above 0°, most seeds were dead after 4 mo; at −15°, 82 percent of seeds germinated after 2 yr.
Barton *(1932)*	Annual and perennial delphinium	−15°	8° and room temp.	−15° gave slightly better results than 8°; sealed seed at −15° retained 50-percent viability for 5 yr.
Beattie and Boxwell *(1939)*	Onion	−6.5°	4.4° and room temp.	Up to 4 yr, 4.4° was better, but for longer periods −6.5° was better.
Kearns and Toole *(1939)*	Chewings fescue	−10°	2°, 10°, 20°, 30°	2 yr was too short to test advantage of −10°; original germination maintained for 2 yr at 10°, 2°, and −10° for 10-percent moisture and 2° and −10° for 14-percent moisture.
Toole and Toole *(1946)*	Soybean	−10°	2°, 10°, 20°, 30°	Nearly full germination maintained for 6 yr at −10° and 18-percent moisture; longevity with 18-percent moisture at higher temperatures was 1 mo to 3 yr; 13.5-percent moisture at 2° and −10° maintained original germination for 10 yr.

TABLE 4.—*Sampling of literature on seed storage at subfreezing temperatures*—Continued

Literature source	Seeds	Storage		Results (°C)
		Below 0° C	Above 0° C	
Pack and Owen *(1950)*	Sugar beet	–23.3° to –12°	None	Seed frozen continuously germinated 83.5 percent in 1928, 81.0 percent in 1949; no reduced vigor during storage.
Barton *(1954)*	5 species of conifers	–4°, –11°, –18°	None	Deteriorated most rapidly at –4°; to maintain viability, –18° was best of 3 temperatures used.
Weibull *(1952, 1955)*	9 vegetable and 16 flower species	–20°	±0° and *ordinary storage* of 5°–20°.	Viability of vegetable seeds retained as long or longer at –15° as at ±0° except parsley; –20° unfavorable for fern asparagus, pansy, parsley, hybrid petunia, and snapdragon; some flower species stored better at –20°, others showed no difference between –20° and *ordinary storage*; results not consistent for some other species.

Sengbusch *(1955)*	Hemp, rye, sugar beet	20°	20°	−20° superior to 20° for hemp and rye, which lost viability after 3 yr; at 20° and −20°, germination was 90 and 98 percent, respectively, after 5 yr; 5 yr too short to evaluate sugar beet as little change in germination at 20° or −20° throughout storage.
Hanscn and Moore *(1959)*	6 species of small-seeded legumes and 2 grasses	−7° to −11°	Laboratory storage	Alfalfa, crimson and white clover, Kobe and Korean lespedeza, *Lespedeza sericea*, orchardgrass, tall fescue. Germination including hard seeds after 3, 4, and 18 yr;[1] −15° superior to lab. storage; germination after 18 yr (2 seed lots per species)—crimson clover and 2 grasses: Lab. 0 percent, subfreezing 39 percent; small-seeded legumes other than crimson clover: Lab. 15.5 percent, subfreezing 79.2 percent.
Barton *(1966b)*	Pyrethrum	−4° and −18.5°	5°, 10°, 20°, 30°	Complete germination for 13 yr, essentially so for 15 yr when stored sealed at −4° and −18.5°; latter better for unsealed seed; lifespan gradually decreased as temperature increased from −18.5° to 30°.

[1] 18-yr data unpublished but supplied by authors.

Weibull (*1952, 1955*) found seeds of a few species that were not benefited by a subfreezing storage temperature. A few other examples have been reported. Roberts *(1972)* analyzed some of them and concluded that, in some instances, statements were made without considering the moisture content of the seed. He believed that low temperature storage is beneficial to maintaining seed viability, but that there may be a few exceptions. This position appears to be supported by the literature.

Early fall freezes while seeds and grain are still in the field have prompted several studies on the deleterious effects of freezing seed and grain at high moisture contents. Kiesselbach and Ratcliff *(1918, 1920)* studied freezing injury to seed corn because it occurred in the fall with considerable frequency in Nebraska. They found that death of seed corn from freezing was directly related to the moisture content of the kernel and to the duration of exposure to low temperatures. Immature kernels with a higher moisture content than mature kernels were more susceptible to freezing injury. All seeds were dead when exposed for 24 hours at the following temperatures and moisture contents:

Temperature (°C)	*Moisture content (percent)*
2.2 to 0	60–65
−4.4 to −6.7	55–60
−6.7 to −8.9	35–40
−13.3 to −15.6	30–35
−17.2 to −20.5	20–25

However, all seeds retained their viability for at least 3 months when stored at −13.3° to −15.6° C and 12- to 14-percent moisture content.

McRostie *(1939)* and Rossman *(1949)* compiled considerable data on freezing injury of corn seed that essentially confirm the findings of Kiesselbach and Ratcliff *(1918, 1920)*. All agree fairly well on the factors causing freeze injury, although their results, when comparable, vary somewhat in detail as would be expected. According to McRostie *(1939)*, 5-day fluctuating temperatures of −17.8° to −23.3° C, −17.8° to −26.1°, and 0° to −26.1° caused more damage to corn seed than a constant exposure to the lowest temperature of any of these regimes. On the other hand, Rossman *(1949)* indicated that repeated freezing and thawing were less injurious than continuous freezing when the total time was the same. He found similar responses to freezing temperatures by the two hybrid cultivars he used. Likewise, Kiesselbach and Ratcliff *(1918, 1920)* found that dent corn and flint corn responded similarly.

In storage experiments with frost-damaged sweet corn seed, Barton *(1960b)* found no significant loss of germination in five lots (four cultivars) over 9 years when stored sealed at −5° C with moisture contents below 13 percent. In open storage at 30° or at −5° in a moisture-saturated atmosphere, the seed decreased in viability by about the same amounts over 7 years.

Apparently seeds of barley, rye, and wheat are nearly as sensitive as corn to cold injury at high moisture contents. Agena *(1961)* found a rapid increase in injury in these species when the grain was stored at −6° C and moisture contents above 22 percent. Under these conditions, injury was evident after 1 day of storage and increased with time. He found that moist grain can be held best for short periods between 0 and −6° but that fungi soon appear under these conditions. Freezing injury was greatest at the radicle tip and secreting area of the scutellum.

Wheat was frozen in liquid nitrogen (bp −195.8° C) for 2 minutes at moisture contents of approximately 19 to 25 percent by Lockett and Luyet *(1951)*. Entire seeds with 19.4-percent moisture content dropped 10 percent in germination because of freezing. Corresponding losses in germination with increase in moisture content of entire seeds were as follows:

Moisture content (percent)	*Germination loss (percent)*
21.2	20
23	89
25.1	100

Imbibed decorticated seeds had a considerably lower moisture content with less germination loss because of freezing than entire seeds imbibed for the same time. The moisture content of imbibed embryos was much greater than that of either entire or decorticated seeds.

Freezing injury to sorghum seed is affected by moisture content of the grain, temperature, and duration of freezing treatment (Robbins and Porter, *1946*; Carlson and Atkins, *1960*; Kantor and Webster, *1967*). In addition, Carlson and Atkins *(1960)* showed that the genotype of the cultivar affects tolerance to freezing injury. Data of Robbins and Porter *(1946)* also suggest genetic effects on freezing tolerance; however, a definite statement to this effect cannot be made since the complete history of each seed lot used is not given. Kantor and Webster *(1967)* used two cultivars in their research but were unable to draw a conclusion with respect to genotype effects. In view of these reports on sorghum seeds and the reports of Kiesselbach and Ratcliff *(1918, 1920)* and of Rossman *(1949)* on corn, the genetic effect on freeze resistance in cereals remains unresolved.

Seed Moisture Content and Relative Humidity

Direct Effect of Moisture on Seed Deterioration

Barton *(1961)* regarded moisture content of utmost importance in seed deterioration. Seed deterioration increases as moisture content is increased.

Boswell et al. *(1940)* illustrated the effects of relative humidity and consequently moisture content of the seeds on their storage life. They stored 10 kinds of vegetable seeds at 2 temperatures, 3 relative humidities per temperature, and under warehouse conditions for 251 days. Samples were tested on 0 day, every 10 days thereafter for 90 days, and after 110 and 251 days of storage. Since 10-, 30-, 50-, 70-, and 90-day tests were not made on peanuts, data were extrapolated to obtain these results for averages.

The averages for 12 test dates times 10 species, or 120 tests per storage condition, are as follows: At 26.6° C, with relative humidities of 78, 66, and 44 percent, the respective average germinations were 55.4, 78.9 and 81.3 percent. At 10°, with relative humidities of 81, 66, and 51 percent, the respective average germinations were 80.7, 82.6, and 82.6 percent. In some respects these results do not reveal the true situations for the following reasons: (1) The storage period was not long enough to measure deterioration of long-storing species, such as tomato or several other species at the more favorable storage conditions, e.g., 10° and 51-percent relative humidity. (2) Averaging species such as onion, with a short storage life, with tomato cancels out the extremes of both species. For example, the percent germinations of these two species on the 251st day of storage at 26.6° and relative humidities of 78, 66, and 44 percent were, respectively, onion, 0.1, 37.4, and 72.9 percent and tomato, 68.1, 85.0, and 91.8 percent.

Since seed deterioration is affected by moisture content, it is important to know what factors affect water absorption and retention as well as their effects. Obviously the thickness, structure, and chemical composition of the seedcoat affect the rate of water absorption and retention by seeds; in hard seeds the seedcoat restricts total water uptake. Of the various seed constituents, proteins are most hygroscopic (readily taking up and holding moisture), carbohydrates are slightly less so, and the lipids are hydrophobic (lacking an affinity for water). Thus, seeds containing relatively high percentages of carbohydrates, proteins, or both, such as rice, other grains, and soybeans, can have moisture contents of about 13–15 percent at 25° C and 75-percent relative humidity, whereas cottonseed, flaxseed, and peanuts, which are rich in oil, would have moisture contents of approximately 9–11 percent at the same temperature and relative humidity (Barton, *1941*). Barton found that seeds of flax, lettuce, onion, peanut, pine, and tomato showed differential water absorption and each species retained its same position relative to other species in all her experiments.

Conceivably a sample of seeds containing both living and dead seeds might take up a different amount of water than a sample of only live seeds or, alternatively, all dead seeds. In studies of this phenomenon by Atkins *(1909)*, Heinrich *(1913)*, Barton *(1941)*, and Simpson and Miller *(1944)*, all agreed that under storage conditions the total moisture uptake and retention do not differ between living and dead seeds. Seeds used included bean, cotton, flax, lettuce, onion, peanut, perennial ryegrass, pine, and sweetpea.

One of Harrington's *(1960d)* so-called Thumb Rules of drying seed stated that for each 1-percent reduction in seed moisture content, the time the seed can be stored without seriously affecting germination is approximately doubled. Harrington *(1972)* also indicated that this rule applies when seed moisture content is between 5 and 14 percent. Roberts *(1972)* developed formulas relating temperature and seed moisture content to the storage life of seeds and indicated that Harrington's *(1960, 1972)* rule agreed with his formulas. However, Roberts *(1972)* pointed out that possibly there are inaccuracies in his own formulas for both seed moisture content and temperature and suggested that the same may be true for Harrington's *(1960, 1972)* rule. On the other hand, both Harrington *(1960, 1972)* and Roberts *(1972)* stated that their procedures give approximate results only.

The doubling of the lifespan of seeds by decreasing the moisture content by 1 percent shows vividly the effects of slight moisture changes, especially at the critical point. In sealed storage the expected life of vegetable seeds with an 8.0-percent moisture content could be doubled by removing 1.0-percent moisture before sealing. On the other hand, certain possible pitfalls should be understood and considered when working near the critical moisture level. Some of these are—

(1) The degree of confidence by which an aliquot of seed tested for moisture represents the lot or bulk.

(2) The degree of accuracy in grinding, drying, and weighing. Usually the air-oven methods are accurate to no more than ±0.1 percent.

(3) The possible variation in seed moisture content among different lots or crops. This can result from differences in chemical composition of the seed caused by different climatic conditions and cultural practices during seed development and maturation.

(4) The hysteresis effect, a phenomenon when at a given relative humidity the equilibrium moisture content of grain and seeds may not always be the same. When seeds lose water and reach equilibrium at any stated relative humidity, the equilibrium moisture content of the seeds may be higher than if dry seeds are allowed to gain moisture to reach equilibrium at the same relative humidity. This means that seeds with a moisture equilibrium value of about 12.5 percent at a relative humidity of 60 percent could possibly have two moisture equilibrium values depending on whether the original seed moisture content was greater or less than 12.5 percent. Hlynka and Robinson *(1954)* showed a difference of about 4.5 percent at a relative humidity of 40 percent for wheat but no difference at 90-percent relative humidity. Hubbard et al.

(1957) found adsorption equilibrium values for corn and wheat to be approximately 1.6 percent higher than desorption values at 33- and 22-percent relative humidity, respectively.

The most common method of determining the moisture content of seeds is to heat them in a forced air oven at a given temperature for a specified time or until constant weight is obtained. The loss of water represents the moisture content of the seeds. The percentage of moisture can be calculated (1) by the wet weight method—the amount of water lost is divided by the initial weight of the sample, and (2) by the dry weight method—the amount of water lost is divided by the dry weight of the sample after drying. Since the original weight in (2) is divided by a smaller number than in (1), the answer will be greater for (2). Thus, in equivalent situations dry weight percentages will be slightly greater than wet weight percentages.

Percentages of moisture based on dry weight are frequently used in research and percentages based on wet weight are usually used for commercial purposes. Seedsmen and seed testing laboratories (Assoc. Off. Seed Anal., Internatl. Seed Testing Assoc., and Soc. Com. Seed Tech.) use the wet weight method to determine seed moisture content. Many U.S. scientists also use this method. However, if the basis for calculating the percentage of moisture is not stated in the literature, it is risky to assume that a specific procedure was followed.

Relationship Between Relative Humidity and Seed Moisture Content

As previously mentioned, the moisture percentages of seeds in equilibrium with a specific relative humidity vary with crop species. At any temperature air will hold a given amount of water in the form of vapor. When the air contains all the moisture it will hold, it is said to be saturated, which is equivalent to the dewpoint or 100-percent relative humidity. Ordinarily the air is not saturated but contains only a fraction of the amount of water it would hold if saturated. Relative humidity reflects the amount of moisture actually in the air as a percentage of the amount of moisture that the air is capable of holding at the same temperature. Thus, relative humidity, expressed as a percentage, is determined as follows: The amount of moisture in the air is divided by the amount of moisture the air is capable of holding at the same temperature and multiplied by 100. Warm air can hold more water than cool air. Thus, if the amount of water in the air is held constant and the temperature is increased, the relative humidity will be decreased. Conversely, if the temperature of the air is lowered, the relative humidity will be increased.

The importance of seed moisture content in retarding seed deterioration cannot be overemphasized. Under all storage conditions the moisture content of the seed will come to equilibrium with the surrounding air if given enough time. In fact, equilibrium is reached between the seeds and the air in the interstitial spaces among the seeds. It has been reached when the net movement of moisture from air to seed, or from seed to air, is zero.

Equilibrium moisture content of seeds is usually determined by the *static method.* By this method seeds are placed in containers to hold the seed sample and acids or dissolved chemicals to provide previously calculated relative humidities. The containers must be absolutely sealed from the outside atmosphere and preferably allow entry to measure the relative humidity of the air. Laboratory desiccators are commonly used for this purpose. The seeds remain in the container with a known air relative humidity until equilibrium is reached, after which the seed moisture content is determined by a reliable method (Hlynka and Robinson, *1954*).

Since the static method may require from a few days to 2–3 months for seeds to reach equilibrium moisture content with the atmosphere, another method that reduces this time interval has been developed. By the *dynamic method* the air is kept in motion and circulated around the sample. In one modification of the procedure, the air is passed through absorption towers, which contain solutions of salts and acids calculated to control the relative humidity surrounding the seeds. In addition, Brewer and Butt *(1950)* found an electric hygrometer to be a useful instrument for determining equilibrium moisture values within 24 to 72 hours. This is basically a desorption method. Seeds of known moisture contents are placed in sealed containers with dry air for a specific time. Thus, moisture moves from the seed to the air in the container. The relative humidity of the air is determined with the electric hygrometer, a hair hygrometer, or other means. According to the authors, their results compared favorably with those obtained by the static method previously described. The electric hygrometer has the disadvantage of requiring seed of a wide range of moisture contents, some below the ambient relative humidities.

Haynes *(1961)* described a vapor pressure method for determining seed hygroscopicity. He plotted hygroscopic curves for several seed species from his data. His curves appear similar to those plotted from data obtained by the static method. The method has several disadvantages and limitations. For these reasons and because it has not been widely accepted, it is not described here.

Hygroscopic equilibrium measurements are usually reported as having been made at 25° C, although exceptions are common. The percentage of moisture the seeds attain at a given relative humidity, usually at 25°, is known as the moisture content equilibrium.

The equilibrium moisture content of several crop species at different relative humidities is given in tables 5–9. In tables 7–8, only limited data were available for forage seeds and most were developed for special conditions. Dexter *(1957)* used relatively high humidities to study the effects of mold formation on equilibrium moisture values for seeds of several forage species, whereas Harrington *(1968)* used low humidities for applications relating to safe sealed storage for at least 3 years. Unfortunately discrepancies exist between Harrington's *(1968)* data at 45-percent relative humidity and Dexter's *(1957)* data at 55-percent relative humidity and to some extent at 65-percent relative humidity. Nevertheless these are the best available tables for forage seeds. Although the data should not be regarded as exact equilibrium moisture values, especially for research purposes, they can be used safely as general guides.

TABLE 5.—*Equilibrium moisture content of vegetable seeds at various relative humidities and approximately 25° C – wet basis*[1]

Vegetable	Moisture content at indicated relative humidity (percent)						
	10	20	30	45	60	75	80
	Percent	*Percent*	*Percent*	*Percent*	*Percent*	*Percent*	*Percent*
Bean, lima	4.6	6.6	7.7	9.2	11.0	13.8	15.0
Bean, snap	3.0	4.8	6.8	9.4	12.0	15.0	16.0
Beet, garden	2.1	4.0	5.8	7.6	9.4	11.2	15.0
Broad bean	4.2	5.8	7.2	9.3	11.1	14.5	17.2
Cabbage	3.2	4.6	5.4	6.4	7.6	9.6	10.0
Cabbage, Chinese	2.4	3.4	4.6	6.3	7.8	9.4	- - -
Carrot	4.5	5.9	6.8	7.9	9.2	11.6	12.5
Celery	5.8	7.0	7.8	9.0	10.4	12.4	13.5
Corn, sweet	3.8	5.8	7.0	9.0	10.6	12.8	14.0
Cucumber	2.6	4.3	5.6	7.1	8.4	10.1	10.2
Eggplant	3.1	4.9	6.3	8.0	9.8	11.9	- - -
Lettuce	2.8	4.2	5.1	5.9	7.1	9.6	10.0
Mustard, leaf	1.8	3.2	4.6	6.3	7.8	9.4	- - -
Okra	3.8	7.2	8.3	10.0	11.2	13.1	14.5
Onion	4.6	6.8	8.0	9.5	11.2	13.4	13.6
Onion, Welsh	3.4	5.1	6.9	9.4	11.8	14.0	- - -
Parsnip	5.0	6.1	7.0	8.2	9.5	11.2	- - -
Pea	5.4	7.3	8.6	10.1	11.9	15.0	15.5
Pepper	2.8	4.5	6.0	7.8	9.2	11.0	12.0
Radish	2.6	3.8	5.1	6.8	8.3	10.2	- - -
Spinach	4.6	6.5	7.8	9.5	11.1	13.2	14.5
Squash, winter	3.0	4.3	5.6	7.4	9.0	10.8	- - -
Tomato	3.2	5.0	6.3	7.8	9.2	11.1	12.0
Turnip	2.6	4.0	5.1	6.3	7.4	9.0	10.0
Watermelon	3.0	4.8	6.1	7.6	8.8	10.4	11.0

[1] Data from Harrington *(1960c)* except last column from Toole *(1942)*.

TABLE 6.—*Adsorbed equilibrium moisture content of grain seeds at various relative humidities and approximately 25° C—wet basis*[1]

Grain	Moisture content at indicated relative humidity (percent)						
	15	30	45	60	75	90	100
	Percent	*Percent*	*Percent*	*Percent*	*Percent*	*Percent*	*Percent*
Barley	6.0	8.4	10.0	12.1	14.4	19.5	26.8
Buckwheat	6.7	9.1	10.8	12.7	15.0	19.1	24.5
Corn shelled, white dent	6.6	8.4	10.4	12.9	14.7	18.9	24.6
Corn shelled, yellow dent	6.4	8.4	10.5	12.9	14.8	19.1	23.8
Oats	5.7	8.0	9.6	11.8	13.8	18.5	24.1
Popcorn, shelled	6.8	8.5	9.8	12.2	13.6	18.3	23.0
Rice	5.9	8.6	10.7	12.8	14.6	18.4	---
Rye	7.0	8.7	10.5	12.2	14.8	20.6	26.7
Sorghum	6.4	8.6	10.5	12.0	15.2	18.8	21.9
Wheat, durum	6.6	8.5	10.0	11.5	14.1	19.3	26.6
Wheat, hard red spring	6.8	8.5	10.1	11.8	14.8	19.7	25.0
Wheat, hard red winter	6.4	8.5	10.5	12.5	14.6	19.7	25.0
Wheat, soft red winter	6.3	8.6	10.6	11.9	14.6	19.7	25.6
Wheat, white	6.7	8.6	9.9	11.8	15.0	19.7	26.3

[1] Adapted from Amer. Soc. Agr. Engin. ASAE ID 245, "Moisture Relations of Grains" (ASAE Handb., *1972*).

The sigmoid shape of a typical curve relating relative humidity to seed moisture content is shown in figure 3. The curve was derived from average moisture contents and relative humidities of seeds of 10 vegetable crop species by Nakamura *(1958)*. Because the seeds of each species have their own moisture-absorbing and moisture-holding potentialities, these curves can be expected to vary to some extent.

Tables 5–9 take into consideration the kinds of seeds and relative humidity, but they have not been calculated to correct for temperature. Most research relating to temperature effects has shown that as temperature is increased when seeds are held at a constant relative humidity, the seed moisture content is decreased. Among the researchers who showed the relationship of temperature to seed moisture equilibrium were Gane *(1941)*, Gay *(1946)*, Henderson *(1952)*, Thompson and Shedd *(1954)*, Hogan and Karon *(1955)*, and Hubbard et al. *(1957)*.

Toole et al. *(1948)* published the seed moisture content of 15 vegetable species stored at 3 relative humidities and 3 temperatures. The results obtained at about 80-percent relative humidity and 11.1°, 21.7°, and 26.7° C permit a comparison of temperature effects. In every instance

TABLE 7.—*Equilibrium moisture content of seeds of grasses and small-seeded legumes at relative humidities below 50 percent and approximately 25° C—wet basis*[1]

Crop	Moisture content at indicated relative humidity (percent)		
	15	30	45
Grasses	*Percent*	*Percent*	*Percent*
Bentgrass, Colonial	6.3	8.2	10.2
Bentgrass, creeping	6.3	8.2	10.2
Bermudagrass, hulled	6.3	8.4	10.5
Bermudagrass, unhulled	6.8	9.3	11.8
Bluegrass, Kentucky	6.2	8.2	10.1
Bluegrass, rough	7.1	9.1	11.0
Bromegrass, smooth	6.6	9.0	11.5
Dallisgrass	5.9	7.8	9.7
Fescue, Chewings	6.5	8.7	10.9
Fescue, creeping red	6.5	8.7	10.9
Fescue, meadow	6.5	8.7	10.9
Fescue, tall	6.5	8.7	10.9
Hardinggrass	6.2	8.1	9.9
Orchardgrass	6.0	8.0	10.0
Redtop	7.0	8.7	10.5
Ryegrass, annual	6.5	8.6	10.7
Ryegrass, perennial	6.5	8.6	10.7
Sorghum, hybrid Ute	6.1	8.3	10.6
Sudangrass, hybrid	5.2	7.3	9.4
Wheatgrass, intermediate	6.1	8.1	10.1
Small-seeded legumes			
Alfalfa	4.5	6.5	8.6
Birdsfoot trefoil, broadleaf	5.9	7.8	9.7
Birdsfoot trefoil, narrowleaf	5.9	7.8	9.7
Clover:			
Alsike	6.1	7.9	9.7
Crimson	5.9	8.0	10.1
Ladino	5.9	7.8	9.7
Red	5.7	7.6	9.4
Rose	5.2	7.2	9.2
Strawberry cv. Palestine	6.6	8.4	10.3
Strawberry cv. Salina	4.3	5.7	7.0
Subterraneum	5.1	6.9	8.7
White	5.4	7.2	9.0
Sweetclover, yellow-blossom	5.9	7.6	9.3

[1] Adapted from Harrington *(1968)*.

the seed moisture content at 21.7° and at 26.7° was lower than at 11.1°. However, without exception the moisture content at 26.7° was higher than at 21.7°. The authors indicated that they had no explanation for

TABLE 8.—*Equilibrium moisture content of seeds of grasses and small-seeded legumes at relative humidities above 50 percent and approximately 23° C—wet basis*[1]

Crop	Moisture content at indicated relative humidity (percent)					
	55	65	70	75	80	85
Grasses	*Percent*	*Percent*	*Percent*	*Percent*	*Percent*	*Percent*
Bentgrass, Colonial	9.8	10.7	11.5	12.5	14.0	14.5
Bluegrass, Kentucky	9.7	10.8	11.3	12.7	14.3	16.4
Bluegrass, rough	9.4	---	11.9	---	13.9	16.2
Bromegrass, smooth	11.0	12.5	13.1	13.7	16.1	18.4
Canarygrass, reed	11.4	12.0	12.5	13.5	14.7	15.7
Fescue, Chewings	10.0	11.2	12.1	---	14.5	16.9
Fescue, creeping red	10.7	11.9	12.6	13.8	15.4	18.0
Fescue, tall	10.5	11.9	12.5	13.2	15.0	17.3
Orchardgrass	9.8	10.5	11.0	12.0	13.4	14.9
Redtop	10.0	10.7	11.0	12.5	13.5	15.0
Ryegrass, annual	11.0	12.1	12.8	14.1	15.7	16.3
Ryegrass, perennial	11.0	12.1	12.8	13.4	14.9	16.6
Sudangrass, piper	10.8	---	11.8	---	14.4	15.6
Timothy	10.9	11.8	12.5	13.6	14.6	16.1
Small-seeded legumes						
Alfalfa	---	---	7.8	9.3	12.5	18.3
Clover:						
Alsike	---	---	9.3	---	15.9	18.9
Ladino	---	---	8.7	10.9	15.4	18.0
Red	---	---	9.1	11.2	15.6	18.7
Crownvetch	---	---	9.4	---	14.2	18.1
Sweetclover, yellow-blossom	---	---	9.3	10.8	12.7	18.3
Trefoil, birdsfoot	---	---	8.3	10.4	13.9	17.2
Vetch, hairy	---	---	11.0	13.0	17.4	18.7

[1] Adapted from Dexter *(1957)*.

this disparity except possibly "the humidity readings in these chambers failed to represent the true humidity surrounding the seeds." The authors pointed out that "the study was not planned to determine the effect of temperature on moisture content of seed, because the recorded mean air humidities may not accurately indicate the true conditions surrounding the seeds." Regardless of the accuracy of the individual values recorded by Toole et al., the data for 15 vegetable species strongly indicate that as temperature is increased, equilibrium moisture content is decreased.

By reducing the results reported by some of these investigators to a common denominator, the following average values were calculated: Average decrease in equilibrium moisture content for each 10° C

TABLE 9.—*Equilibrium moisture content of seeds of several crops at various relative humidities and approximately 12° to 25° C*[1]

Crop	Moisture content at indicated relative humidity (percent)									
	10	20	30	40	50	60	70	80	90	100
	Percent	*Percent*	*Percent*	*Percent*	*Percent*	*Percent*	*Percent*	*Percent*	*Percent*	*Percent*
Broadbean	4.7	6.8	8.5	10.1	11.6	13.1	14.8	17.2	22.6	+27.5
Cotton	3.7	5.2	6.3	6.9	7.8	9.1	10.1	12.9	19.6	- - -
Flax	3.3	4.9	5.6	6.1	6.8	7.9	9.3	11.4	15.2	21.4
Lupine	4.2	6.2	7.8	9.1	10.5	11.7	13.4	16.7	+25.0	- - -
Pea	5.3	7.0	8.6	10.3	11.9	13.5	15.0	17.1	22.0	26.0
Peanut	3.0	3.9	4.2	5.1	5.9	7.0	8.5	11.1	17.2	- - -
Rape	3.1	3.9	4.5	5.2	6.0	6.9	8.0	9.3	12.1	+15.5
Soybean	- - -	5.5	6.5	7.1	8.0	9.3	11.5	14.8	18.8	- - -
Sugar beet	4.4	6.3	8.0	9.4	10.7	12.0	13.3	16.6	20.5	22.5

[1] Compiled from Hall *(1957)*, Karon and Hillery *(1949)*, Kreyger *(in* Owen, *1956)*, and Simpson and Miller *(1944)*.

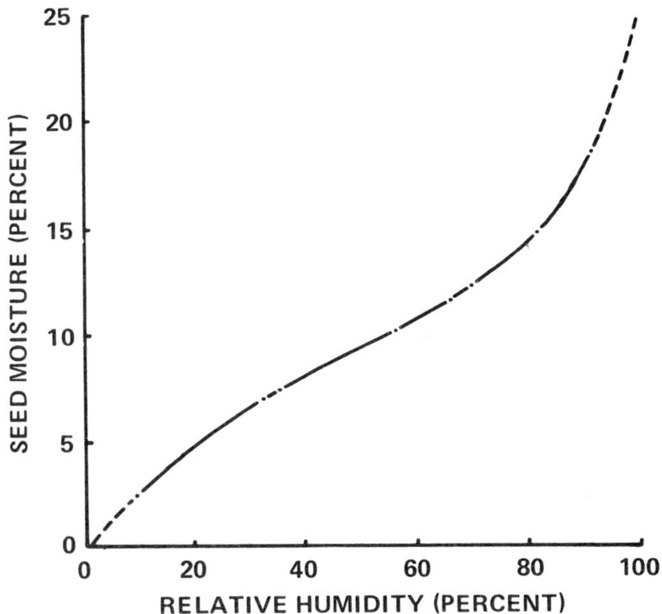

FIGURE 3.—Relationship of atmospheric relative humidity and seed moisture content of vegetable seeds. (Courtesy of Nakamura, *1958.*)

increase in temperature—*wheat* 0.65 (Gane, *1941*), 0.43 (Gay, *1946*), *wheat and corn* 0.72 (Thompson and Shedd, *1954*), 0.73 (Hubbard et al., *1957*), *blue lupine* 0.90, *corn* 1.4, *crimson clover* 0.62, and *sorghum* 0.3 (Haynes, *1961*). Using the data of Toole et al. *(1948)* for 15 species of vegetable seeds stored at approximately 80-percent relative humidity, the average difference per 10° calculated for storage temperatures of 11.1° and 26.7° amounted to 0.21-percent moisture content. For most of these crops these relative humidity values do not change greatly as the temperature is increased or decreased. Although the moisture content of the seeds increases with an increase in relative humidity, these values are not greatly affected by temperature.

The slight effect of temperature on seed moisture content is illustrated in figure 4 for sorghum and wheat. Although the curves for these two species are different, the lines for the different temperatures are parallel, indicating a continuing relationship as humidity is increased. These relationships do not hold true in all instances. Curves for corn and rescuegrass seed (fig. 4) show how the curves can deviate as humidity is increased. For corn, the equilibrium moisture content decreases 0.96 percent for each 10° C increase at 30-percent relative humidity, but at 70-percent relative humidity it decreases 2.3 percent for each 10° increase. At a relative humidity of about 56 percent the seed moisture content of rescuegrass seed is the same at −1° and at 49°. As the

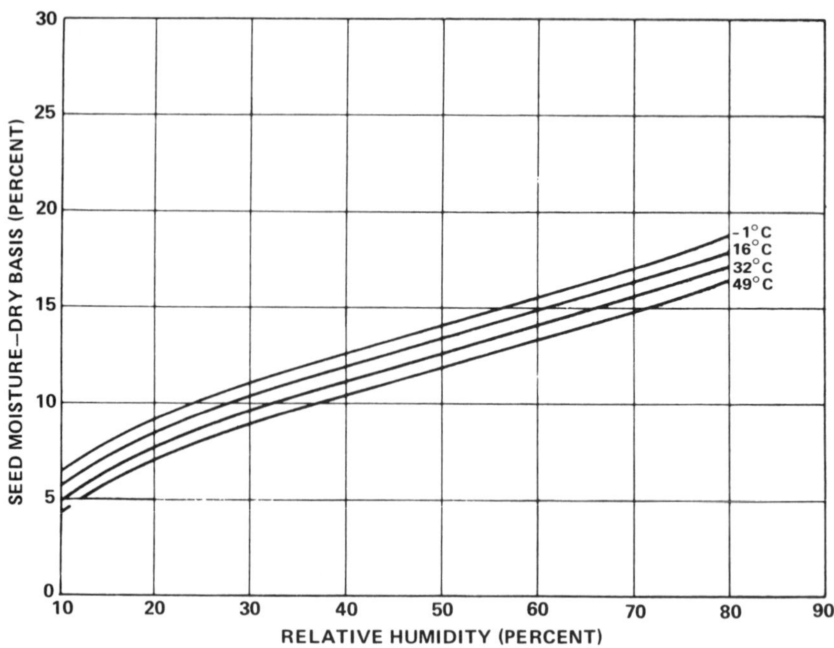

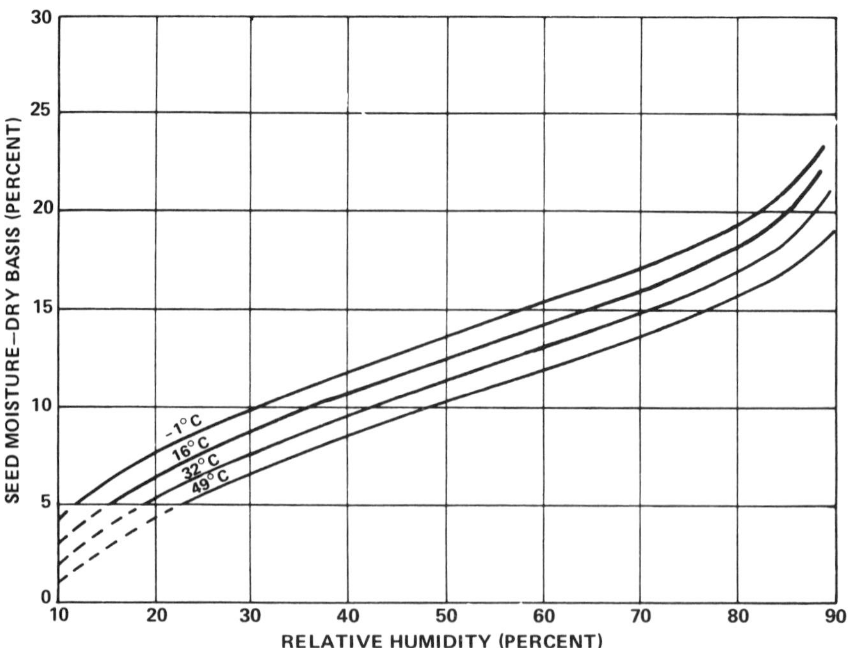

FIGURE 4.—Equilibrium relative humidity curves for sorghum, wheat, corn, and rescuegrass *(top to bottom)*. (Courtesy of Haynes, *1961.*)

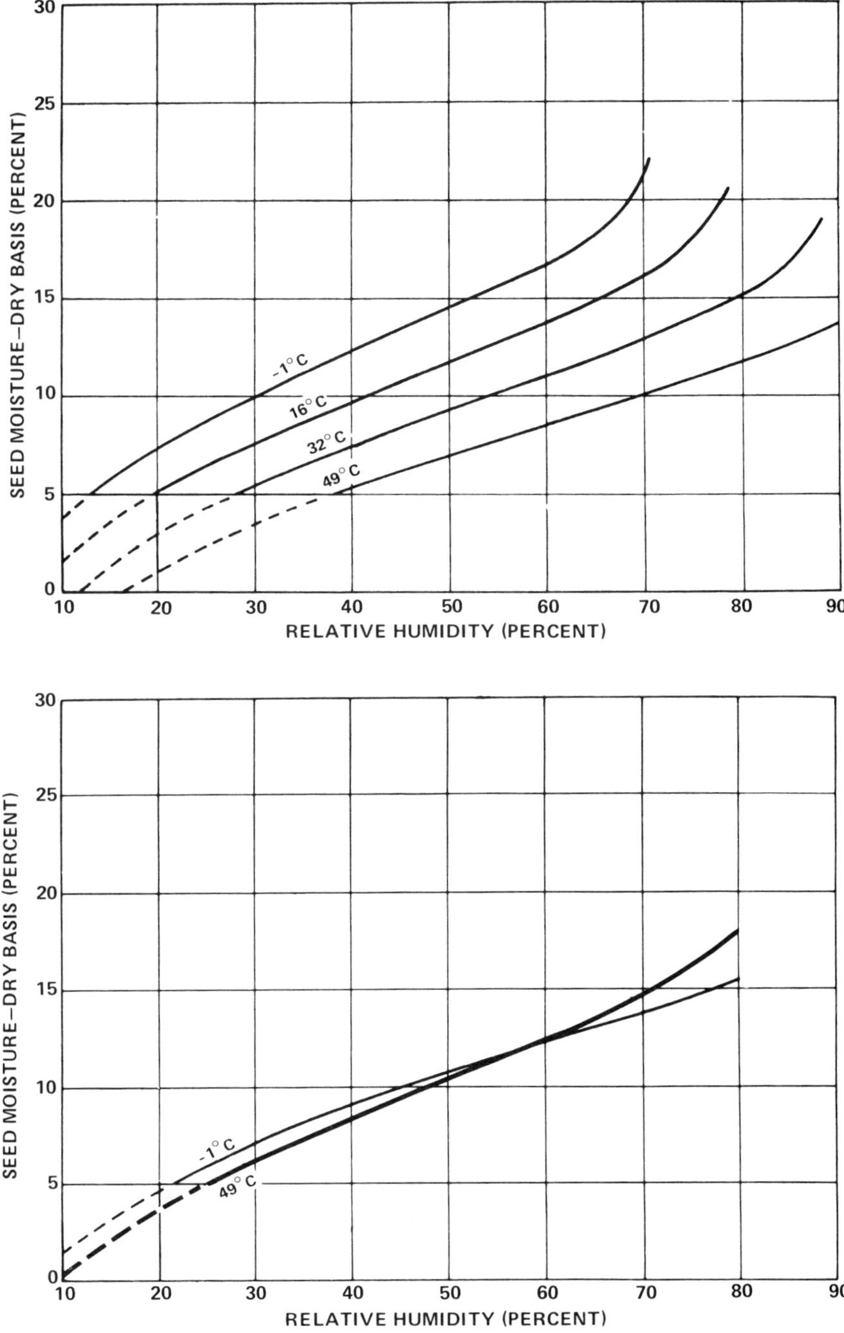

FIGURE 4.—Equilibrium relative humidity curves for sorghum, wheat, corn, and rescuegrass *(top to bottom)* (Courtesy of Haynes, *1961.*)—Continued.

relative humidity is decreased below about 54 percent, the seeds at 49° hold less water than seeds at −1°. The difference at 25-percent relative humidity is approximately 0.2-percent moisture for each 10°. On the other hand, as the relative humidity is increased above 56 percent, the seeds hold more water at 49° than at −1°. The difference at 80-percent relative humidity amounts to approximately 0.44-percent moisture for each 10° change.

Rate of Moisture Absorption and Movement

Studies on moisture absorption and movement in grains and seeds fall into three categories: (1) Laboratory studies on small samples, (2) studies on bulk samples of a few pounds to several tons, and (3) studies on moisture movement of mixed grain with different moisture contents. For information on moisture movement in grain, see Fisher and Jones *(1939)*.

Much of the research on equilibrium moisture in small seed samples has been concerned with ascertaining the time required for the seed to reach equilibrium. This is regulated by the time required for moisture (1) to penetrate the seedcoat and (2) to transfer within the seed, based on the assumption that the relative humidity surrounding the seeds is uniformly distributed. Since seeds of different species differ as to seedcoat permeability to moisture, as well as constituents of the endosperm and embryo, it is only logical that the time required to reach equilibrium moisture will vary with the species.

Dillman *(1930)* found that dry flaxseeds absorbed hygroscopic moisture more rapidly than wheat seeds, which absorbed moisture more rapidly than alfalfa seeds. Likewise, Pixton and Warburton *(1968)* reported similar results with the soft wheat cultivar Cappelle, which reached 90-percent equilibrium moisture content in less time than did the hard wheat cultivar Manitoba.

Temperature also affects the rate of water absorption. Dillman *(1930)* found the absorption by dry corn, flax, and wheat seed to be twice as rapid at 30° C as at 10° but no change from 30° to 40°. Whitehead and Gastler *(1946–47)* determined that at 20- to 80-percent relative humidity, sorghum and wheat seeds reached equilibrium moisture content within 15 days at 20°, whereas nearly 70 days were required at 1°.

The difference between the vapor pressure of the seeds and that of the surrounding atmosphere is an extremely important factor in determining the rate of moisture movement. For example, if rice of 13-percent moisture content is placed in an atmosphere of 95-percent relative humidity, it will increase in moisture content faster than at 70-percent relative humidity. Likewise, the rice will lose moisture more rapidly if placed at 20- than at 50-percent relative humidity.

Pixton and Warburton *(1968)* measured the rate of hygroscopic moisture absorption of two cultivars of wheat at relative humidities from 35 to 90 percent. They found that the time required to reach equilibrium moisture was neither a gradual progression from 35- to 90-percent relative humidity (absorption) nor a gradual degression from 90 to 35 percent (desorption). Rather, equilibrium was reached in the following order of relative humidity percentages, with the first number being the fastest rate:

Absorption—35, 45, 80–90, 60–80
Desorption—85–90, 60–80, 35, 45

Moisture uptake is rather rapid for the first 2 to 3 days for many kinds of seeds, after which it decreases. According to Babbitt *(1949)*, wheat, whole or with pericarp and seedcoat removed, reached near equilibrium moisture content within 40 hours and the increase thereafter was slow. Results reported by Pixton and Warburton *(1968)* for wheat more or less confirmed Babbitt's report. They found that 90 percent of equilibrium moisture content was reached within 5–14 days by absorption and 2–9 days by desorption. On the other hand, Breese *(1955)* indicated that at a relative humidity from 10 to 90 percent, rough rice required at least 60 days to reach equilibrium moisture. Small samples of chaffy grass seeds appeared to come to equilibrium moisture content within 3–5 days over a wide range of relative humidity (Dexter, *1957*). Harrington and Aguirre *(1963)* dried seeds of six vegetable species to near 0-percent moisture content and placed them in a saturated atmosphere. After 64 hours they had regained moisture to reach the following moisture percentages (wet weight basis): Onion 9.4, carrot 9.4, cabbage 8.3, tomato 8.1, lettuce 8.0, and muskmelon 7.5.

Caution should be exercised in placing much confidence in data obtained at high relative humidity, especially when observed over long periods of time. Under these conditions, the changes in weight of samples may be affected not only by changes in water content of the seeds but also by the growth of micro-organisms and by respiration.

Movement of moisture into or out of bulk grain or seed is very slow. The time for moisture to enter and be distributed within a seed is only a small fraction of the time required for it to move through a mass of seed. Instead of the moisture moving from seed to seed or from grain to grain, it penetrates the mass through the interstitial airspaces among the seeds. The movement has been measured. Equilibrium was reached in wheat at a depth 1 inch below the surface after 400 hours (Babbitt, *1949*). Seeds of the six species of vegetables used by Harrington and Aguirre *(1963)*, previously discussed, had an average moisture content of 8.5 at the surface of the bulk after 64 hours but only 0.7 percent 3 inches below the surface. Frey *(1960)* found seeds of wheat and birdsfoot trefoil at the center of 100–kg bags to contain 2–4 percent more moisture than seeds taken from the surface just inside the walls of the jute bags. The seeds had been stored at five temperature-relative humidity conditions prior to sampling.

These data show how slowly seeds and grain at the center of large bulks come to equilibrium moisture content. Were this not so, large bulks of grain within or outside of enclosures could not be stored safely. Although the relative humidity surrounding a bin of grain may change gradually or abruptly, the outer layers of the grain insulate the inner layers to a degree. Each layer forms a temporary barrier, which mediates the change of relative humidity in the next layer. With continuing high humidity, the barrier is pushed farther toward the center of the bulk until equilibrium is reached in the entire lot.

A difference in temperature between two areas or zones of a storage bin or silo can cause seed or grain to deteriorate. When warm air moves to a colder zone and becomes chilled to the dewpoint, the moisture in the air is deposited on the seeds as liquid. This situation can exist when seeds lie in a zone between a cold outside wall of a storage structure and the interior core where the temperature is higher. Christensen *(1970)* demonstrated this principle by using grain sorghum with a moisture content of 14.3 percent. A temperature difference of 12° – 14° over a distance of 6–8 inches resulted in rapid transfer of moisture from the warmer to the cooler part of the grain. The grain where moisture accumulated became heavily infested with storage fungi and decayed. Equilibrium moisture content was attained within 12 days.

Effect of Extreme Desiccation on Viability and Vigor

Since increased knowledge and improved technology have promoted the drying of seeds to very low moisture levels for storage, shipment, or both in sealed containers, it is important to know whether the drying process causes losses in viability and vigor. There are essentially no reports of seed damage caused by drying to approximately 6-percent moisture content, but several workers have reported damage when seeds were dried to 5 percent or lower. Toole and Toole *(1946)* reported an increase in abnormalities of soybean seedlings grown from seeds stored for 3 months at a moisture content slightly above 5 percent. Abnormalities consisted primarily of deep cracks across the cotyledons that interfered with using the stored food reserves.

Roberts *(1959)* found that timothy seed retained its viability better in storage as its moisture content was decreased to about 7 percent, but storage at 5 percent decreased viability. Ching et al. *(1959)* found a similar trend with perennial ryegrass seeds. At 6-percent moisture content they did not retain their viability as well as at 8.3 percent.

Apparently caution must be exercised in storing seeds of some cultivars of cotton, sorghum, and various other crop species at very low seed moisture contents. Phillis and Mason *(1945)* reported that as the relative humidity of the storage atmosphere was decreased below 40 percent, germination capacity decreased after storage for 5 years at 24°–27° C. Seeds stored at 40-percent relative humidity germinated 75 percent, whereas seeds at 30-, 20-, and 0-percent relative humidity germinated 70, 65, and 22 percent, respectively. Seed moisture content ranged from 6.9, to 5.2, to near 0 at 40-, 20-, and 0-percent relative humidity, respectively. Phillis and Mason *(1945)* found that by placing the low moisture content seeds at a slightly increased relative humidity, thus gradually increasing seed moisture content, they could overcome the deleterious effects of desiccation.

Nutile *(1964a, 1964b)* found that sorghum seeds that had been dried to 3.0- to 3.5-percent moisture content showed delayed germination and produced seedlings with damaged radicles, which subsequently developed many secondary roots. However, he was able to overcome the apparent injury by placing the dry seeds at 55-percent relative humidity until their moisture content increased to 11 percent. Such preconditioned seeds germinated normally.

McCollum *(1953)*, Pollock et al. *(1969)*, and Pollock and Manalo *(1970)* noted that cotyledon cracking in garden bean seeds can be caused by stresses during imbibition and germination as well as by mechanical means. Very dry bean seeds that developed high percentages of cotyledon cracking with rapid imbibition developed less cracking when allowed to gradually absorb moisture to about 12 percent before imbibition.

Much of the damage previously considered as desiccation and mechanical damage appears really to be stress damage. If so, seeds can be dried to very low moisture levels for safe sealed storage, but sensitive kinds may require special handling prior to planting. Seeds dried to a very low moisture content for sealed storage apparently must be allowed to equilibrate with normal atmospheric conditions before planting except in very arid areas, where special slow imbibition techniques will have to be employed.

Seeds of some species that can be dried safely to 2- to 3-percent moisture content are injured when dried to about 1 percent or lower. At such low moisture content the first indication of injury is reduced rate of germination, followed in storage experiments by decreased germination. With severe drying, the symptoms of injury may appear immediately after drying. As the critical point of damage is approached but not reached, the symptoms will be apparent after storage. These generalizations are based on data by Harrington and Crocker *(1918)*. Seeds of Kentucky bluegrass reduced to 1.5-percent moisture content suffered no loss either in germination capacity or in vigor. With 0.2-percent moisture content no loss in germination was evident, but vigor was reduced considerably, whereas at 0.1-percent moisture content germination capacity had dropped by 5 percent and vigor was seriously reduced. Further reduction of the seed moisture content reduced the germination percentage by one-half and all seedlings were very weak. This trend has been found for seeds of several crop species but not all. Effect of extreme desiccation on seed viability and vigor is shown in table 10.

Went and Munz *(1949)* subjected seeds of 113 plant species native to California to extreme desiccation over phosphorus anhydride (P_20_5) and sealed them under vacuum in small glass tubes. No tube was opened until the appointed time for testing. After 1 year of storage, five species had lost three-fourths of their germination capacity and some others had declined to a less extent. It is not known whether this loss was due to desiccation or to some other factor.

Based on current information, seeds of most crop species can be dried to 2- to 3-percent moisture content without significant injury provided injury is not caused by another factor associated with the drying process, such as high temperature. However, drying below 3- to 4-percent moisture content is not recommended for storage of seeds in commerce, because extremely dry seeds may be damaged by too rapid rehydration when planted.

Symptoms of desiccation injury may not be apparent until after storage. They include cracked cotyledons, damage to food transport system in the embryo, stunted radicle with unusually heavy development of secondary roots, stubby primary root and shoot, protrusion of radicle without further development, and decreased germination compared with nondamaged seeds. The last symptom is usually an excellent indication of reduced vigor.

Another precaution to observe when drying seed is the possibility of inducing dormancy. Nutile and Woodstock *(1967)* induced 10- to 25-percent "low-temperature" dormancy in seeds of six sorghum cultivars by drying. When placed to germinate at 15° C, dormancy was apparent, but at 25° the blocks to germination did not exist. Species, and possibly some cultivars, for which experimental data are not available might well be treated as suspect.

Interrelationship of Temperature, Seed Moisture Content, and Storage Life

To some degree both temperature and seed moisture content can be controlled for practical storage. The nature and degree of control will depend on the economics applicable to specific situations. These decisions may be affected by local climate and the quantity of seed to be stored as well as by the length of storage.

Available data on storage of hempseeds demonstrate the effects of

TABLE 10.—*Effect of extreme desiccation on seed viability and vigor of several crop species*

Source	Crop	Seed moisture content	Comment
		Percent	
Harrington and Crocker *(1918)*	Kentucky bluegrass	Below 1	Discussed in text.
	Barley	0.7–0.8	No injury from drying.
	Sudangrass	0.5–0.6	Do.
	Johnsongrass	0.1–0.2	Slight decrease in germination.
	Wheat	About 1.0	No injury from drying.
Green *(1961)*	Corn (maize)	Below 2.0	Below 2-percent moisture content, viability reduced in direct relationship to length of drying period.
Smith and Gane *(1938)*	Chewings fescue	Approximately 0.5	Drying over phosphorus anhydride for 25 days did not reduce germination significantly.
Evans *(1957b)*	Perennial ryegrass	0.66	No immediate loss in viability, but loss in vigor was apparent.
Nutile *(1964b)*	Kentucky bluegrass	Below 1	Severe injury after 5 years of storage.
	Red fescue	Below 1	Do.
	Highland bentgrass	Below 1	Little or no injury after 5 years of storage.
	Onion	Below 1	Do.
	Cabbage	Below 1	Do.
	Cucumber	Below 1	Do.
	Lettuce	Below 1	Do.

TABLE 10.—*Effect of extreme desiccation on seed viability and vigor of several crop species*—Con.

Source	Crop	Seed moisture content	Comment
		Percent	
Nutile *(1964b)*—Con.			
	Carrot	Below 1	Medium injury after 5 years of storage.
	Tomato	Below 1	Do.
	Celery	Below 1	Severe injury after 5 years of storage.
	Eggplant	Below 1	Do.
	Pepper	Below 1	Do.
Joseph *(1929)*	Parsnip	0.4	No indication of injury.
Went and Munz *(1949)*	Native plants of California.	---	Discussed in text.

temperature and moisture on storage life. In 1950, seed lot 2 was grown in Kentucky and seed lot 4 in Maryland, and in 1951 they were stored at Beltsville, Md., until 1960–61, when they were transferred to the National Seed Storage Laboratory, Fort Collins, Colo. The tests over the first 5½ years were made at Beltsville by Toole et al. *(1960)* and those from 9 to 12 years of storage by Clark et al. *(1963).* Lot 2 represented healthy seed of strong vigor, which had a germination capacity of 98 percent when stored. The viability of lot 4 when stored was approximately 88 percent. There was no significant loss in germination of lot 2 stored in sealed containers at 6.2-percent moisture. Unfortunately the data are not available for seed stored at 6.2-percent moisture and 21° C beyond 5½ years.

Table 11 shows the effect of seed moisture content, temperature, and length of storage on seed germination. Seed lot 2 stored at 6.2-percent moisture content and 10° C germinated 96 percent after 12 years, whereas seed stored at 9.5-percent moisture content and 10° lost its viability completely during the same period. Seed stored at 9.5-percent moisture content and 10° required 11 years to reach 5-percent germination, but seed stored at the same moisture content and 21° required only 2 years to reach this germination level. There is greater variation among germination values for lot 4. This is common in seed lots of declining viability or of low vigor. Otherwise the trends for moisture and temperature effects are comparable for the two seed lots.

The effect on an oilseed crop, soybean, of a combination of high temperature (30° C) and high moisture content (18.1 percent for Mammoth Yellow and 17.9 percent for Otootan) is given in table 12. Only 14 percent of the Mammoth Yellow seeds and none of the Otootan seeds were viable after 1 month of storage. The high moisture content of both cultivars stored at −10° caused little or no deterioration until after 72 months of storage. These data show that the viability of soybean is maintained longer when stored at subfreezing temperatures than at 2°. No seeds of either cultivar with 9.4- and 8.1-percent moisture content decreased in viability when stored at −10°, 2°, or 10° for 10 years (cols. 2–4).

The results of an experiment with nine species of vegetable seeds and peanut for storage periods of 110 and 250 days are given in table 13. This table differs from tables 11 and 12 in that it is based primarily on vegetable seeds, evaluates storage conditions by crop species and by their averages, and concentrates on intermediate temperatures. Although the moisture contents at which the seeds were stored are not shown in table 13, they can be estimated for the different crop species at each humidity level by referring to tables 5 (veg.) and 9 (peanuts).

The average results for both storage periods show that as storage temperature or storage relative humidity is increased the germination

TABLE 11.—*Germination of 2 lots of hempseed stored in sealed containers at 8 temperature-moisture conditions for 12 years*[1]

Storage period (years)	Germination at indicated seed moisture content and temperature (°C)							
	6.2 percent				9.5 percent			
	−10	0	10	21	−10	0	10	21
	Percent	*Percent*	*Percent*	*Percent*	*Percent*	*Percent*	*Percent*	*Percent*
				SEED LOT 2				
0	98	98	98	98	98	98	98	98
½	98	96	97	98	99	98	98	97
1	96	99	97	98	98	95	94	64
2	96	96	97	97	98	97	95	5
3	93	97	98	95	98	95	93	0
4	97	98	98	97	98	98	91	0
5	98	96	97	93	98	99	84	0
5½	93	96	98	98	97	96	65	0
9	---	96	93	---	---	97	13	---
10	92	94	96	---	95	91	6	---
11	93	94	96	---	94	83	5	---
12	97	98	96	---	97	95	0	---
				SEED LOT 4				
0	87	87	87	87	89	89	89	89
½	87	80	87	85	84	87	92	85
1	83	80	92	93	78	87	92	85
2	85	79	90	87	87	82	91	10
3	88	85	85	86	76	83	76	0
4	84	88	93	86	90	89	77	0
5	84	82	86	86	87	78	60	0
5½	80	84	73	75	86	80	42	0
9	---	82	86	---	---	77	3	---
10	53	73	71	---	58	51	1	---
11	66	79	68	---	63	50	1	---
12	64	69	66	---	79	62	0	---

[1] Data from Toole et al. *(1960)* and Clark et al. *(1963)*.

of the seeds is reduced. After 110 days the averages of seeds stored at 10° C and 81-percent relative humidity differed by only 0.6 percent from those at 26.7° and 44-percent relative humidity. However, after 250 days the averages differed by 4.1 percent. The data for 44- and 78-percent relative humidity at 26.7° show that seeds of tomato retained good germination longest and seeds of onion and peanut the shortest time.

Barton *(1939)* in experiments with aster seeds over 3 years reported on seed deterioration under several storage conditions. One of the most important results in table 14 is the consistently high germination of the seeds stored under controlled conditions of low temperature and low to moderate seed moisture content compared to the relatively rapid deterioration of seeds stored at ambient temperature and the same moisture content.

Vacuum and Gas Storage

For many years research has been conducted on the effects of a partial vacuum and such gases as carbon dioxide, oxygen, and nitrogen on the longevity of various kinds of seeds. The reported results from these studies are variable and in some instances appear contradictory. This confusion undoubtedly results from the widely divergent test methods employed by the various researchers. Because of the lack of complete information on the test procedures used in numerous studies, direct comparisons cannot be made between and among the various data found in the literature. Likewise sealed storage cannot be directly compared with open storage, because in sealed containers oxygen concentration in the atmosphere decreases and the carbon dioxide concentration increases with time (Harrington, *1963*), whereas in open storage the composition of the atmosphere remains constant. Because it is not feasible to continually adjust the composition of the atmosphere in sealed containers, most studies have not included gas analyses.

Some workers paid no attention to either seed moisture or storage temperature, whereas others attempted to control one or the other. Several used air-dry seeds and room temperature, both of which provide a minimum of information about the conditions actually used. Regardless of the kind of seed, the air-dry moisture content in one geographic area is not necessarily the same as that in another. Likewise room temperature is not consistent from place to place nor from month to month, or even from day to day in most localities. To accurately assess the protective value of either a partial vacuum or a gas, all environmental conditions have to be considered. Reports on a few kinds of seeds will illustrate the variable results in the literature.

Barley

The higher the oxygen content of the storage environment, the shorter the viability period for barley seeds (Roberts and Abdalla, *1968*). The deleterious effects of oxygen are produced at relatively low oxygen concentrations and are most pronounced at the higher seed moisture levels.

Corn

Since the seedsman wants to store his seeds at the highest practical moisture level and temperature, Goodsell et al. *(1955)* stored seed of two corn hybrids at approximately 8-, 10-, 12- and 14-percent moisture sealed in air, carbon dioxide, and nitrogen at approximately −18°, 4°, 16°, and 29° C. Generally the average pooled germination of the two

TABLE 12.—*Seed germination of 2 soybean cultivars stored at 3 moisture contents and 5 temperatures for 10 years*[1]

Storage period (months)	Germination at indicated moisture content and temperature (°C)														
	Reduced moisture[2]					Natural moisture[3]					High moisture[4]				
	−10	2	10	20	30	−10	2	10	20	30	−10	2	10	20	30
	Percent	*Percent*	*Percent*	*Percent*	*Percent*	*Percent*	*Percent*	*Percent*	*Percent*	*Percent*	*Percent*	*Percent*	*Percent*	*Percent*	*Percent*
							MAMMOTH YELLOW[5]								
1	---	---	---	---	---	---	---	97	99	98	---	---	97	93	14
3	98	98	96	96	97	96	97	96	99	87	98	94	91	96	0
5	96	94	93	97	96	96	94	95	98	0	97	98	96	85	---
9	91	89	89	99	95	93	94	98	97	---	94	96	94	0	---
12	93	93	95	99	87	93	95	98	93	---	96	97	88	---	---
24	96	94	98	96	0	96	97	96	0	---	99	97	1	---	---
48	94	96	99	89	---	98	96	88	---	---	95	81	---	---	---
72	99	97	96	70	---	97	98	39	---	---	89	0	---	---	---
96	95	94	93	47	---	98	95	19	---	---	70	---	---	---	---
120	92	95	94	0	---	98	90	0	---	---	17	---	---	---	---
							OTOOTAN[5]								
1	---	---	---	---	---	---	---	94	93	92	---	---	91	92	0
3	92	93	88	90	88	94	90	95	94	64	94	94	95	79	---
5	92	90	94	94	89	91	94	92	93	18	94	90	89	0	---
9	93	89	90	96	92	93	91	94	88	0	94	92	89	---	---
12	94	94	88	95	91	89	93	92	73	---	96	92	76	---	---
24	90	93	88	92	74	93	95	93	0	---	92	88	0	---	---
48	89	95	95	88	1	90	93	85	---	---	83	30	---	---	---
72	91	98	93	86	0	94	92	15	---	---	93	0	---	---	---
96	94	96	94	79	---	98	94	0	---	---	27	---	---	---	---
120	94	93	95	68	---	98	91	---	---	---	44	---	---	---	---

[1] Data from Toole and Toole *(1946)*.
[2] Mammoth Yellow 9.4 and Otootan 8.1 percent.
[3] Mammoth Yellow 13.9 and Otootan 13.4 percent.
[4] Mammoth Yellow 18.1 and Otootan 17.9 percent.
[5] Mammoth Yellow 97- and Otootan 93-percent germination before storage.

TABLE 13.—*Seed germination of 10 crops after 110 and 250 days of storage at 6 temperature-relative humidity conditions*[1]

Crop	Germination before storage	Germination at indicated temperature and relative humidity[2]					
		10° C			26.7° C		
		51 percent	66 percent	81 percent	44 percent	66 percent	78 percent
	Percent	*Percent*	*Percent*	*Percent*	*Percent*	*Percent*	*Percent*
			AFTER 110 DAYS' STORAGE				
Bean, kidney	97	88	88	83	80	81	60
Bean, lima	76	68	60	75	59	65	54
Beet	83	87	87	85	93	88	78
Cabbage	93	93	71	89	90	90	66
Carrot	93	90	90	88	89	89	56
Corn, sweet	82	62	66	58	66	67	13
Onion	80	74	75	61	79	65	0
Peanut	83	[3]75	71	76	60	43	0
Spinach	73	80	72	73	76	69	23
Tomato	92	93	90	90	92	87	77
Average	85.0	81.0	79.0	77.8	78.4	74.4	42.7

See footnotes at end of table.

TABLE 13.—*Seed germination of 10 crops after 110 and 250 days of storage at 6 temperature-relative humidity conditions*[1]—Continued

Crop	Germination before storage	Germination at indicated temperature and relative humidity[2]					
		10° C			26.7° C		
		51 percent	66 percent	81 percent	44 percent	66 percent	78 percent
	Percent	*Percent*	*Percent*	*Percent*	*Percent*	*Percent*	*Percent*
		AFTER 250 DAYS' STORAGE					
Bean, kidney	97	78	79	91	92	87	0
Bean, lima	76	58	69	61	47	61	26
Beet	83	85	81	87	88	79	9
Cabbage	93	92	92	88	91	89	1
Carrot	93	91	88	88	90	87	1
Corn, sweet	82	79	60	57	70	66	0
Onion	80	75	76	64	73	37	0
Peanut	83	70	68	49	58	29	0
Spinach	73	73	71	63	75	63	0
Tomato	92	91	91	87	92	85	68
Average	85.0	78.2	77.5	73.5	77.6	68.3	10.5

[1] Data from Boswell et al. *(1940)*.

[2] Percent germination adjusted to nearest whole number.

[3] Correction made in original data based on 77-percent germination after 2½ months' storage and 70 percent after 8⅓ months' storage.

TABLE 14.—*Germination of aster seeds after storage at 12 temperature-moisture conditions over 36 months*[1]

Months	Temperature	Germination at indicated seed moisture content (percent)			
		4.6	6.7	7.9	Ambient
	°C	*Percent*	*Percent*	*Percent*	*Percent*
6	−5	90	88	85	87
	5	87	88	80	72
	Room	89	84	85	86
12	−5	87	88	87	86
	5	91	90	91	64
	Room	90	83	83	83
18	−5	85	79	87	82
	5	83	80	84	46
	Room	78	78	60	71
24	−5	66	71	68	72
	5	68	71	56	4
	Room	64	35	0	3
30	−5	84	86	85	82
	5	84	87	87	0
	Room	75	33	0	15
36	−5	90	92	91	86
	5	91	93	89	0
	Room	76	9	0	0

[1]Data from Barton *(1939)*. Seeds stored at room temperatures were sealed in glass tubes; all others were sealed in tin cans; average germination at start of experiment assumed to be 90–92 percent.

hybrids in the three storage gases was still 95 percent or better after 5 years for the seeds at 8-, 10-, and 12-percent moisture at −18° and 4° and those containing 8- and 10-percent moisture at 16°. Seeds containing 12- or 14-percent moisture at 16° and those with 8- and 10-percent moisture at 29° deteriorated rapidly after 1 year. Corn seeds containing 12- to 14-percent moisture were practically all dead after one-half to 1 year, regardless of the surrounding gas.

Sayre *(1940)* stored corn seeds with 18-percent moisture in oxygen, carbon dioxide, and nitrogen at 30° C. The seeds in oxygen died within 3 years and the germination of the seeds in carbon dioxide and nitrogen dropped noticeably. At low temperatures corn seeds with 18-percent moisture sealed in carbon dioxide and nitrogen had good germination for 5 years.

Struve[8] dried corn seeds to near 0-percent moisture, sealed them in oxygen and in nitrogen, and stored them at −30° to 50° C. Seeds at −30°

[8]STRUVE, W. M. DRYING AND GERMINABILITY OF MAIZE. 1958. [Unpublished Ph.D. thesis. Copy on file Dept. of Botany, Iowa State Univ., Ames.]

remained unchanged through 31 months, those at 50° in nitrogen showed a progressive reduction in vigor, and seeds in oxygen died within 7 months. He also stored corn seeds with 1-percent moisture content in atmospheres containing 0-, 20-, 60-, and 100-percent oxygen at 0°, 5°, 10°, 15°, 25°, 35°, and 50°. He concluded that oxygen concentration may become an important factor in the deterioration of corn seeds in sealed storage.

Flower Seeds

Primula sinensis sealed in carbon dioxide declined only 30 percent in viability over a 7-year period, whereas unsealed seeds lost all viability (Lewis, *1953*). Seeds of *Salvia splendens* deteriorated seriously when sealed under a vacuum (Chopinet, *1952*). Aster seeds kept equally well when sealed in air or a partial vacuum (Barton, *1939*). Sealing under a partial vacuum had no advantage over sealing in air for maintaining the viability of verbena seeds (Barton, *1939*).

Barton *(1960a)* stored seeds of *Lobelia cardinalis* with 6.9- and 4.7-percent moisture in sealed glass vials in air, oxygen, carbon dioxide, nitrogen, and vacuum and in open containers for up to 25 years at laboratory temperature, 5°, and −5° C. In all cases, viability was lost more rapidly at laboratory temperature than at 5° and −5° and at 6.9-percent than at 4.7-percent moisture. At laboratory temperature, 6.9-percent moisture seeds in air, open or sealed, lost viability rather rapidly. Seeds sealed in oxygen lost viability even more rapidly, whereas carbon dioxide, nitrogen, and a partial vacuum extended the life of the seeds for 6 to 8 years. When seed moisture was reduced to 4.7 percent, seed longevity was increased at laboratory temperature. Carbon dioxide, nitrogen, and vacuum were superior to air and oxygen, with oxygen causing the most rapid deterioration. There were no significant differences in the response of seeds to atmospheres of carbon dioxide, nitrogen, or under vacuum at room temperature or to air, oxygen, carbon dioxide, nitrogen, or vacuum at 5° and −5° except those resulting from increased time in storage.

Grasses

The effect of nitrogen was negligible on the longevity of Chewings fescue (Gane, *1948a*) and meadow fescue (Evans et al., *1958*). There was no advantage in using either nitrogen or a partial vacuum rather than air for sealed storage of seeds of Kentucky bluegrass and creeping red fescue (Isely and Bass, *1960*). In fact, when the seeds were subjected to an unfavorable temperature, loss of viability was more rapid for seeds packaged with nitrogen or under vacuum than with air.

Legumes

Seeds of both red and white clover stored under vacuum and in nitrogen were shorter lived than those stored unsealed (Davies, *1956*). Red clover seeds containing 10.3-percent moisture when sealed with carbon dioxide lost all viability in 23 years, but when calcium chloride was used with the carbon dioxide, only about one-third of the initial viability was lost (Evans, *1957a*).

No atmosphere tested, including air, vacuum, carbon dioxide, nitrogen, helium, and argon, was consistently or significantly better than all others for 2 years of sealed storage of crimson clover seeds (Bass et al., *1963a*).

Oilseeds

Soybeans in open storage for nearly 6 years lost viability (Guillaumin, *1928*), whereas seeds sealed in an atmosphere free of oxygen germinated 92 percent and those under a vacuum had 100-percent viability.

Low moisture (7 percent) cottonseeds retained their initial viability when sealed in air, oxygen, carbon dioxide, or nitrogen and stored at 21° and 32° C (Simpson, *1953*). Seeds with 13-percent moisture dropped to one-half to two-thirds of the original germination under all storage conditions. The loss of germination with oxygen was no greater than with carbon dioxide or nitrogen; however, the loss in air was greater than in the pure gases.

Bass et al. *(1963b)* found that air, vacuum, carbon dioxide, nitrogen, helium, or argon was neither consistently nor significantly better than the others for sealed storage of safflower and sesame seeds for 2 years.

Rice

Much of the literature pertains to the storage of rice seeds for both food and seed. In areas where rice is grown, seed moisture content tends to remain high even when the seeds are air-dry. Deterioration of high moisture (20.8 percent) seeds at 30° C can be delayed for a few weeks by sealing them in an atmosphere of carbon dioxide mixed with 1,000 p/m of ethylene oxide (Kaloyereas, *1955*).

Kondo and Okamura *(1927, 1929, 1930, 1934)* and Kondo et al. *(1929)* found that both rough and hulled rice can be stored sealed with carbon dioxide or air for up to 4 years with little loss of viability provided the seed moisture content is less than 13 percent. They reported that carbon dioxide had a slight advantage over air. Rice dried to 5-percent moisture and sealed in an atmosphere of nitrogen germinated 99 percent after 8 years, but with 13-percent moisture all viability was lost (Sampietro, *1931*). Seeds with either 5- or 13-percent moisture lost all viability when sealed in carbon dioxide, air, or under a partial vacuum.

Sorghum

Sorghum seeds during the second year of storage retained significantly higher germination when sealed under a partial vacuum than when sealed in air, carbon dioxide, nitrogen, argon, or helium (Bass et al., *1963a*).

Vegetable Seeds

For pea seeds the period of safe storage decreased as the oxygen concentration in the storage atmosphere increased from 0 to 21 percent (Roberts and Abdalla, *1968*). The deleterious effects of oxygen were more pronounced at the higher seed moisture contents. There was no advantage in using sealed storage under nitrogen or a partial vacuum rather than air for onion seeds (Gane, *1948b*; Isely and Bass, *1960*). However, seeds sealed in carbon dioxide retained their viability better than did similar seeds sealed in air (Lewis, *1953*; Harrison and McLeish, *1954*; Harrison, *1956*). Vacuum, carbon dioxide, nitrogen, helium, and argon storage had no advantage over sealed-in-air storage for lettuce seeds during 2 years (Bass et al., *1962*). Although lettuce seeds sealed in carbon dioxide retained their viability better at room temperature than did similar seeds sealed in air, storage in carbon dioxide revealed differences in longevity between cultivars (Harrison and McLeish, *1954*; Harrison, *1956*).

Cabbage seeds stored equally well when sealed in air, nitrogen, or a partial vacuum (Isely and Bass, *1960*). Parsnip seeds retained their viability better when sealed in carbon dioxide than in air (Lewis, *1953*; Harrison, *1956*). Dandelion seeds retained their viability about equally well when sealed in a partial vacuum or in air, except seeds containing 7.9-percent moisture seemed to deteriorate more rapidly in a partial vacuum (Barton, *1939*).

Wheat

Wheat, like rice, is an important food crop, which has received much attention because the condition of storage greatly affects its milling and processing qualities as well as its seed germination. To improve the storage of wheat seeds for both planting and food purposes, studies have been made on the effects of nitrogen and carbon dioxide on seed viability. Loss of viability of high moisture wheat seeds can be delayed for several days by sealing under either 50 or 75 percent of carbon dioxide (Peterson et al., *1956*) or nitrogen (Glass et al., *1959*). However, once deterioration starts, it proceeds rapidly. It can be delayed for several additional weeks by combining nitrogen storage with a lower temperature (20° C); however, even this combination is unsatisfactory for extended storage.

Confirmation Studies at the National Seed Storage Laboratory

Obviously some kinds of seeds under certain circumstances are benefited by vacuum or gas storage. The question then is "Under what conditions is vacuum or gas storage practical and desirable?" To more fully understand the interrelationship involved, a comprehensive study was undertaken at the National Seed Storage Laboratory, Fort Collins, Colo. One lot each of the following kinds of seeds was adjusted to 4-, 7-, and 10-percent moisture and sealed in air, under vacuum, and in carbon dioxide, nitrogen, helium, and argon and stored at −12°, −1°, 10°, 21°, and 32° C: 'Dixie' crimson clover, 'Great Lakes' lettuce, 'Pacific No. 1' safflower, 'Margo' sesame, and 'RS 610' hybrid sorghum. No gas analyses were made after storage. Germination tests were made at the time of storage and at intervals thereafter. Because commercially available equipment was very expensive, Bass and James *(1961)* developed a simple, inexpensive device for vacuum and gas sealing of tin cans. The complete sealer (fig. 5) consists of a vacuum chamber, soldering gun, three-way valve, vacuum gage, several lengths of air hose, hose clamps, a brass tee fitting, and a source for both vacuum and gas.

Germination tests were made in electronically controlled water-curtain germinators operated at 20° C for crimson clover and lettuce and a 20° to 30° night-to-day alternation for safflower, sesame, and sorghum. Each test consisted of two 100-seed replicates planted according to official procedures (Association of Official Seed Analysts, *1954*) except safflower, for which official procedures had not been developed at that time. Safflower was planted the same as sorghum. In all tests only normal seedlings—those capable of producing plants—were recorded. For crimson clover, which has hard seeds, the percentage of hard seeds was added to the percentage of normal seedlings to give total germination.

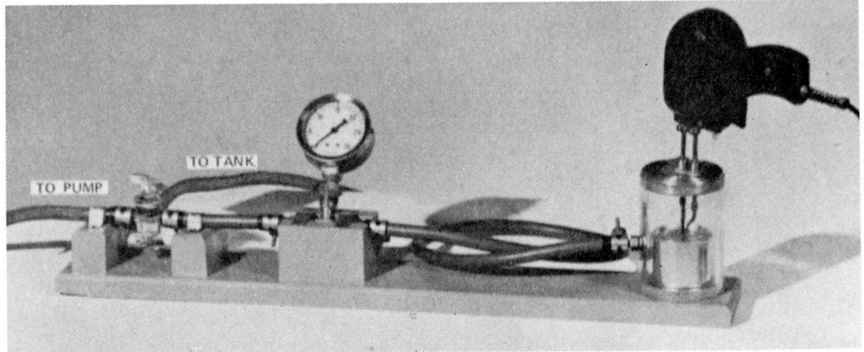

PN–5396

FIGURE 5.—Device for sealing tin cans from which air has been extracted by vacuum or gas has been added.

The data (table 15) show that regardless of the kind of seed, no storage atmosphere consistently gave the highest germination percentage at all temperatures for seeds of all moisture levels tested. The data also show that there are distinct differences between kinds of seed in their response to temperature and seed moisture content.

Of the five kinds of seeds, sorghum showed the least sensitivity to the interaction between seed moisture content, storage temperature, and storage atmosphere (table 15). Only the 10-percent moisture seed showed really drastic germination reductions when stored at 32° C. At that temperature seeds in all atmospheres except those under a partial vacuum (3 percent) and in argon (5 percent) were completely dead. At 21°, seeds with 10-percent moisture in a partial vacuum, helium, and argon did not show a significant reduction in germination, but seeds in all other atmospheres did. Significant reductions in germination were recorded for 4-percent moisture seeds under vacuum at 21° and in air, nitrogen, and helium at 32°. Significant reductions in germination occurred for 7-percent moisture seeds in all atmospheres at 32°. No significant reductions were recorded for seeds of any moisture content in any atmosphere at 10°, −1°, and −12°.

Crimson clover seeds (table 15) stored in paper envelopes (check) showed a significant decrease in germination after 8 years of storage at each temperature for each initial moisture level except 4- and 10-percent moisture seeds held at −12° C. Except for 7- and 10-percent moisture seeds at 10°, 21° and 32°, only an occasional sample of sealed seeds showed a significant decline in germination. Such declines were not consistent for storage temperature, seed moisture, or surrounding atmosphere. A significant decline in germination of sealed seeds stored at −12° was recorded for 7-percent moisture seeds in a partial vacuum and for 10-percent moisture seeds in air. At −1°, significant germination losses occurred with 7-percent moisture seeds held in nitrogen and with 10-percent moisture seeds held in air, vacuum, carbon dioxide, nitrogen, and helium. The significant germination declines recorded at 10° occurred for 7-percent moisture seeds in air, carbon dioxide, and nitrogen and for 10-percent moisture seeds in air, vacuum, carbon dioxide, nitrogen, and helium. At 21°, significant decreases occurred for 7-percent moisture seeds in air, carbon dioxide, nitrogen, and helium and for all 10-percent moisture seeds.

The only significant decrease in germination of 4-percent moisture seeds held at 32° C was for those in nitrogen. All 7-percent moisture seeds showed a significant decline, with the highest germination maintained in an atmosphere of argon. Argon also gave the highest germination of 4-percent moisture seeds at 32°.

The sealed 10-percent moisture seeds at 32° C showed a significant decline in germination the first year of storage and germinated 12

TABLE 15.—*Germination of 5 kinds of seeds stored with approximately 4-, 7-, and 10-percent moisture content at 5 temperatures in various atmospheres in sealed metal cans; check samples were in paper envelopes*

Temperature (°C)	Seed moisture	Initial germination	Germination[1] after 8 years of storage in indicated atmosphere						
			Air	Vacuum	Carbon dioxide	Nitrogen	Helium	Argon	Check
	Percent	*Percent*	*Percent*	*Percent*	*Percent*	*Percent*	*Percent*	*Percent*	*Percent*
				CRIMSON CLOVER					
−12	4	82	84	86	87	86	89	86	78
	7	92	87	80	89	89	83	85	78
	10	91	80	89	86	82	85	84	83
−1	4	82	85	88	84	91	80	86	66
	7	92	83	83	84	81	85	83	71
	10	91	70	69	81	81	74	85	75
10	4	82	82	78	82	78	83	88	44
	7	92	76	86	78	79	85	86	41
	10	91	42	77	74	80	81	85	45
21	4	82	80	85	82	76	88	75	33
	7	92	81	89	80	81	82	87	34
	10	91	5	11	15	11	7	5	34
32	4	82	80	77	77	72	80	80	71
	7	92	69	74	70	76	76	79	69
	10	91	0	0	0	0	0	0	73
				LETTUCE					
−12	4	97	96	96	94	94	96	94	92
	7	97	98	97	94	95	97	96	96
	10	95	94	95	93	93	92	95	95
−1	4	97	90	95	89	90	93	93	84
	7	97	76	62	96	96	95	95	90
	10	95	91	91	92	96	93	94	96

See footnote at end of table.

TABLE 15.—*Germination of 5 kinds of seeds stored with approximately 4-, 7-, and 10-percent moisture content at 5 temperatures in various atmospheres in sealed metal cans; check samples were in paper envelopes*—Con.

Temperature (°C)	Seed moisture	Initial germination	Germination[1] after 8 years of storage in indicated atmosphere						
			Air	Vacuum	Carbon dioxide	Nitrogen	Helium	Argon	Check
	Percent	*Percent*	*Percent*	*Percent*	*Percent*	*Percent*	*Percent*	*Percent*	*Percent*
				LETTUCE—con.					
10	4	97	90	5	93	95	89	95	0
	7	97	1	91	93	94	90	92	0
	10	95	0	0	0	0	0	0	0
21	4	97	90	90	92	89	90	93	0
	7	97	0	0	0	0	0	0	0
	10	95	0	0	0	0	0	0	0
32	4	97	36	90	92	85	92	88	2
	7	97	0	0	0	0	0	0	0
	10	95	0	0	0	0	0	0	0
				SAFFLOWER					
−12	4	95	87	89	92	92	90	89	86
	7	94	83	87	82	89	86	83	88
	10	95	79	75	79	85	72	81	88
−1	4	95	91	92	89	93	86	95	90
	7	94	90	87	89	84	87	84	86
	10	95	72	78	79	70	79	69	88
10	4	95	89	84	91	88	89	89	81
	7	94	85	75	75	80	81	81	88
	10	95	0	0	0	0	0	0	86
21	4	95	90	67	90	87	90	93	64
	7	94	0	0	0	0	0	0	61
	10	95	0	0	0	0	0	0	54

32	4	95	89	88	90	86	86	88	53
	7	94	0	0	0	0	0	0	60
	10	95	0	0	0	0	0	0	53
SESAME									
−12	4	94	92	90	94	94	93	89	92
	7	92	89	90	91	87	88	87	93
	10	88	0	0	0	0	0	0	89
−1	4	94	92	89	91	96	89	93	94
	7	92	87	85	87	84	87	84	90
	10	88	0	0	0	0	0	0	90
10	4	94	90	92	92	92	89	94	88
	7	92	0	0	0	0	2	1	95
	10	88	0	0	0	0	0	0	84
21	4	94	91	91	93	89	89	94	88
	7	92	0	0	0	0	0	0	88
	10	88	0	0	0	0	0	0	83
32	4	94	87	84	89	86	87	87	86
	7	92	0	0	0	0	0	0	88
	10	88	0	0	0	0	0	0	70
SORGHUM									
−12	4	92	90	92	89	92	88	92	94
	7	95	94	93	91	92	95	86	95
	10	91	94	92	93	95	93	92	92
−1	4	92	92	92	92	89	92	90	95
	7	95	94	91	93	89	93	91	89
	10	91	87	95	92	90	93	86	93
10	4	92	90	92	92	93	91	91	71
	7	95	90	87	87	89	88	90	76
	10	91	92	83	85	83	85	90	76

See footnote at end of table.

TABLE 15.—*Germination of 5 kinds of seeds stored with approximately 4-, 7-, and 10-percent moisture content at 5 temperatures in various atmospheres in sealed metal cans; check samples were in paper envelopes*—Con.

Temperature (°C)	Seed moisture	Initial germination	Germination[1] after 8 years of storage in indicated atmosphere						
			Air	Vacuum	Carbon dioxide	Nitrogen	Helium	Argon	Check
	Percent	*Percent*	*Percent*	*Percent*	*Percent*	*Percent*	*Percent*	*Percent*	*Percent*
				SORGHUM—con.					
21	4	92	83	82	93	84	90	87	55
	7	95	86	87	88	87	87	85	84
	10	91	72	82	70	75	82	87	80
32	4	92	65	89	82	69	81	90	50
	7	95	65	77	70	71	72	74	46
	10	91	0	3	0	0	0	5	40

[1] Least significant difference at 5-percent level of probability is 10 percent.

percent or less at the end of the second year. By the end of the eighth year, even the few hard seeds initially present were dead.

Maximum germinations were distributed among atmospheres as follows:

Atmosphere and combinations (number)	*Seed moisture (percent)*	*Temperature (°C)*
Air (1)	4	32
Vacuum (3)	7	10
		21
	10	−12
Carbon dioxide (2)	7	−12
	10	21
Nitrogen (2)	4	−1
	7	−12
Helium (4)	4	−12
		21
		32
	7	−1
Argon (6)	4	10
		32
	7	10
		32
	10	−1
		10

In some tests the same germination was recorded for seeds in two or more atmospheres. This finding applies to the tabular data for lettuce, safflower, sesame, and sorghum (pp. 72–74).

The data for lettuce (table 15) show that seed moisture and storage temperature had more effect on germination than did the composition of the surrounding atmosphere. Drying to 4-percent moisture before sealing made possible the storing of lettuce seeds at as high as 32° C with significant differences between atmospheres evident only at 32°. The apparent significant decline to 5-percent germination for 4-percent moisture seeds in a partial vacuum at 10° resulted from a defective seal after evacuation. The 1-percent germination for 7-percent moisture seeds in air at 10° also resulted from a defective seal. In both tests a fine hole in the solder allowed the seeds to absorb moisture from the atmosphere until an unsafe level was reached. These data emphasize the need for extreme care in sealing seed containers. Highest germination of lettuce seeds in sealed cans occurred as follows:

Atmosphere and combinations (number)	*Seed moisture (percent)*	*Temperature (°C)*
Air (2)	4	−12
	7	−12
Vacuum (3)	4	−12
		−1
	10	−12
Carbon dioxide (2)	4	32
	7	−1
Nitrogen (4)	4	10
	7	−1
		10
	10	−1
Helium (2)	4	−12
		32
Argon (3)	4	10
		21
	10	−12

Safflower seeds must be dried to 4-percent moisture (table 15) for safe, sealed storage at ordinary temperatures regardless of the surrounding atmosphere. Seeds containing 7-percent moisture cannot be stored safely above 10° C, and 10-percent moisture seeds cannot be stored above −1°. All sealed 10-percent moisture safflower seeds showed a significant loss in germination during 8 years of storage. The array of highest germinations for safflower seeds sealed in various atmospheres was as follows:

Atmosphere and combinations (number)	*Seed moisture (percent)*	*Temperature (°C)*
Air (2)	7	−1
		10
Vacuum (0)	---	---
Carbon dioxide (4)	4	−12
		10
		32
	10	−1
Nitrogen (3)	4	−12
	7	−12
	10	−12

Atmosphere and combinations (number)	Seed moisture (percent)	Temperature (°C)
Helium (1)	10	−1
Argon (2)	4	−1
		21

The data for sesame seeds (table 15) show that this oilseed does not store well when sealed in any atmosphere at moisture levels above 7 percent unless the temperature is kept at −1° C or lower. For sealed storage in any atmosphere at higher temperatures seed moisture content must be reduced to 4 percent. The distribution among atmospheres of the highest germination percentages for sesame seeds was as follows:

Atmosphere and combinations (number)	Seed moisture (percent)	Temperature (°C)
Air (1)	7	−1
Vacuum (0)	---	---
Carbon dioxide (4)	4	−12
		32
	7	−12
		−1
Nitrogen (2)	4	−12
		−1
Helium (2)	7	−1
		10
Argon (2)	4	10
		21

No atmosphere was consistently better than all others for preserving the germination of sealed sorghum seeds. Therefore it is interesting to note that among seed moistures and storage temperatures the following combinations, with atmospheres that gave the highest germinations after 8 years of sealed storage, were about equally divided:

Atmosphere and combinations (number)	Seed moisture (percent)	Temperature (°C)
Air (4)	4	−1
	7	−1
		10
	10	10

Atmosphere and combinations (number)	*Seed moisture (percent)*	*Temperature (°C)*
Vacuum (4)	4	−12
		−1
	7	32
	10	−1
Carbon dioxide (3)	4	−1
		21
	7	21
Nitrogen (3)	4	−12
		10
	10	−12
Helium (2)	4	−1
	7	−12
Argon (5)	4	−12
		32
	7	10
	10	21
		32

The distribution of highest germinations for all five kinds of seeds among the various atmospheres in sealed metal cans is summarized as follows:

Atmosphere, combinations (number), and kind of seed	*Seed moisture (percent)*	*Temperature (°C)*
Air (10):		
Crimson clover	4	32
Lettuce	4	−12
	7	−12
Safflower	7	−1
		10
Sesame	7	−1
Sorghum	4	−1
	7	−1
		10
	10	10
Vacuum (10):		
Crimson clover	7	10
		21
	10	−12

Atmosphere, combinations (number), and kind of seed	*Seed moisture (percent)*	*Temperature (°C)*
Vacuum (10)—Continued		
Lettuce	4	−12
		−1
	10	−12
Sorghum	4	−12
		−1
	7	32
	10	−1
Carbon dioxide (11):		
Safflower	4	−12
		10
		32
	10	−1
Sesame	4	−12
		32
	7	−12
		−1
Sorghum	4	−1
		21
	7	21
Nitrogen (14):		
Crimson clover	4	−1
	7	−12
Lettuce	4	10
	7	−1
		10
	10	−1
Safflower	4	−12
	7	−12
	10	−12
Sesame	4	−12
		−1
Sorghum	4	−12
		10
	10	−12
Helium (11):		
Crimson clover	4	−12
		21
		32
	7	−1

Atmosphere, combinations (number), and kind of seed	*Seed moisture (percent)*	*Temperature (°C)*
Helium (11)—Continued		
Lettuce	4	−12
		32
Safflower	10	−1
Sesame	7	−1
		10
Sorghum	4	−1
	7	−12
Argon (18):		
Crimson clover	4	10
		32
	7	10
		32
	10	−1
		10
Lettuce	4	10
		21
	10	−12
Safflower	4	−1
		21
Sesame	4	10
		21
Sorghum	4	−12
		32
	7	10
	10	21
		32

It is obvious from this distribution of highest germinations that no one atmosphere is decidedly better than the others for safe, sealed storage of seeds.

The data in table 15 show that regardless of the kind of seeds or the atmosphere in the sealed container, only adequately dried seeds (4-percent moisture) retained their germination reasonably well during 8 years of storage at 32° C. Some kinds did not keep well at 21°, and sesame seeds with 7- and 10-percent moisture lost all viability at 10° except for 1 or 2 percent when sealed in helium or argon.

It is obvious from the data that seed moisture content at the time of sealing has a far greater effect on seed longevity than does the surrounding atmosphere. For most situations the added expense of using an atmosphere other than air appears unnecessary. However, even for sealing in air, most kinds of seeds require additional drying before sealing if the sealed seeds are to be subjected to a temperature above 10° C. In general, the added advantage, if any, of an atmosphere other than air is not realized except during very long-term storage.

Illumination

Only limited studies have been made on the possible effects of light on stored seeds. At least four investigators, who have studied the potential effects of white light on the storage life of seeds, have found no positive effect of illumination that could not be explained by the reduction in seed moisture content caused by the light treatment (Tammes, *1900*; Bacchi, *1955*; Wyttenbach, *1955*; Litynski and Urbaniak, *1958*). Harrington *(1972)* indicated that ultraviolet rays from the sun can reduce longevity of seeds before harvest and hasten deterioration of stored seeds. Since Harrington *(1972)* did not indicate the source of his information, these statements are assumed to be extrapolations from the effects of ultraviolet radiation on other biological materials.

Barton *(1947)* claimed that when *Cinchona ledgeriana* seeds with a moisture content as high as 9.4 percent were exposed to light during storage in a laboratory, there was some evidence of harmful effects. However, the author indicated that the moisture content of the control samples stored in darkness may have been reduced during storage and thus provided a more suitable storage environment. Also, seeds stored in sealed glass tubes and not resealed for future use after opening did not confirm the initial conclusion.

Perhaps the most comprehensive study of this type was made by Litynski and Urbaniak *(1958)*. For 4 years they studied light effects on the storage life of seeds of eight vegetable species and seven field crop species that had been exposed to white light, orange-yellow light, blue-violet light, diffused light of reduced intensity, and darkness. In studying possible benefits of spectral characteristics of the filters used and light intensity, they found no consistent benefit. Any apparent benefit might be explained by the possible drying of the seeds by the orange-yellow rays as opposed to the lack of drying by the blue-violet rays. Diffused light was not beneficial to the retention of seed viability.

Jensen *(1941, 1942)* claimed that the storage life of cabbage and cauliflower seeds was increased significantly by illumination simultaneously with a mercury-quartz lamp and a heat lamp. Principal irradiation by the mercury-quartz lamp was from 300μ to 450μ with a maximum energy peak at about 325μ, whereas the heat lamp emitted radiation from about 350μ to 550μ with a peak at 485–550μ. In Jensen's work *(1941, 1942)* the seeds were exposed to light in the open, either on a canvas with constant stirring or on a moving belt provided with ventilation. Although we cannot disagree with the claims that light affects the storage life of seeds, we believe that the author's findings did not exclude the possibility of seed dehydration, which resulted from the treatment procedure responsible for the increased storage life.

Harrington *(1972)* explained the relationship of a macromolecular water layer, which insulates a macromolecule against oxidation and subsequent deterioration of molecules, cells, and embryonic tissue. The removal of this macromolecular layer renders the molecule vulnerable to oxidation by several agents including ultraviolet light. Maes and Bauwen *(1951)* studied the effect of infrared rays on preserving wheat seeds and some oilseeds. They found that infrared rays inhibited the lipolytic degradation of wheat germ and some oilseeds. The phenomenon was not caused by a change in the moisture content of the germ, as the moisture content remained unchanged during radiation and storage.

The published information on the effects of illumination on the storage life of seeds is controversial. Until these controversies are resolved, no general recommendation can be made that would apply to field, vegetable, and flower seeds.

Respiration and Heating

Respiration has been defined as an oxidative-reduction process occurring in all living cells, whereby chemical action occurs, producing compounds and releasing energy that are partly used in various life processes. Respiration is discussed here only as it relates to practical seed storage. The following phases of respiration may be considered: (1) Depletion of food reserves, (2) formation of intermediate or end products that may affect seed storage, and (3) release of energy, much of which is in the form of heat. Of these three phases, perhaps the release of heat has the greatest application to seed storage.

Depletion of Food Reserves

Since respiration is an oxidation process, there must be a substrate (the seed) with which the oxygen can combine. Also, respiration is possible only in the presence of enzymes, some of which are very specific and some more general in their functions. As respiration proceeds, more and more of the seed food reserves are used up. There is some evidence that the carbohydrate reserves in the endosperm of wheat, on which much research of this nature has been conducted, are used before the reserves in the embryo. Ordinarily the sugars and starches are digested before the lipids and proteins. When stored seeds are protected from insects and vermin and the environment is unfavorable for the development of molds and bacteria, the respiration process continues under some conditions to the state or condition wherein the seed appears as charcoal, for example, carbonized barley seeds taken from the Anatolian excavations of Alishar, dating to about 1000 to 3000 B.C. It is doubtful that seeds stored under reasonably favorable conditions of temperature and humidity would lose enough dry weight over 5 to 10 years to significantly weaken them.

Respiration Products That Retard This Process

One product of respiration is carbon dioxide. Most relevant research has shown that in a closed system the accumulation of carbon dioxide inhibits respiration (Crocker and Barton, *1953*). In bag and bulk storage the carbon dioxide is usually dissipated and thus has little effect on respiration. Milner and Geddes (*in* Anderson and Alcock, *1954, pp. 152–220*) indicated that renewal of oxygen in open storage can occur with normal changes in barometric pressure. The inhibition of respiration by carbon dioxide in a closed system can be used advantageously by storing seeds in sealed containers in which the carbon dioxide replaces the air. Inhibitory concentrations of carbon dioxide in respiring soybeans were found to be from 12 to 14 percent. Milner and Geddes (*in* Anderson and Alcock, *1954, pp. 152–220*) reported that a carbon dioxide concentration of 7 percent affected respiration in wheat and at 12 percent inhibition was marked.

Release of Heat in Respiration

The typical textbook equation for respiration is

$$C_6H_{12}0_6 + 6CO_2 = 6CO_2 + 6H_20 + 673 \text{ kg calories.}$$

This equation shows that for oxidation of 1 mole of a hexose, 6 moles of oxygen are required; that 6 moles each of carbon dioxide and water result from this oxidation; and that 673 kg calories of energy are released. The complete combustion of a gram of glucose yields 3.76 calories. Not all glucose or other sugars proceed to complete combustion. Moreover the amount of heat released per unit of substrate varies with the kind of material involved. The important point here is that heat is evolved from respiration.

Under favorable storage conditions the heat of respiration is of little or no concern for practical seed storage. However, at higher moisture levels the heat of respiration can produce much damage to stored seed. This would probably occur with freshly harvested seed having a high moisture content (De Witt et al., *1962*), with seed in storage areas not well protected from the weather, with seed in the holds of ships where the temperatures are high, or where it is not adequately protected from the elements.

Measurement of Respiration

Because respiration in the resting seed proceeds very slowly, equipment and methods to measure the amount and rate must be precise.

Methods commonly used to study respiration in seeds include measuring oxygen absorption and carbon dioxide evolution. Because each method has weaknesses (see Bartholomew and Loomis, *1967*), many researchers measure both oxygen absorption and carbon dioxide evolution, which also provide data for computing the respiration quotient. This quotient is the ratio of carbon dioxide produced to the amount of oxygen consumed, usually written as $CO_2/0_2$. Attempts have been made to measure water production in respiration, decrease in dry matter (substrate), evolution of heat, and loss of energy as a measure of respiration. These methods have not been as successful for measuring small bulk samples as measurement of oxygen absorption and carbon dioxide evolution.

Some Factors Affecting Respiration

The apparent respiration of seed lots and the associated heat may arise not only from the seeds but also from the fungi and bacteria on the seeds, especially at seed moisture contents in equilibrium with approximately 75-percent relative humidity and higher. According to Milner and Geddes (*in* Anderson and Alcock, *1954, pp. 152–220*), critical moisture percentages that increase respiration are wheat 14.6, barley 14.5, corn 14.2, and flaxseed 11.0. Other examples of equilibrium moisture content of seeds are in tables 5–9. Since mold spores do not germinate at relative humidity levels much below 75 percent, the relative humidity of the storage area should be considerably less than 75 percent for safe storage for more than 1 year. Molds grow readily at humidities above 75 percent. Bacteria are of minor importance in seed storage as their relative humidity requirement is above 90 percent. If held at optimum temperature (usually 25° to 30° C), fungus spores may germinate at a lower relative humidity than at a nonoptimum temperature.

Respiration increases as temperature is increased until limited or terminated by inactivation of enzymes or enzyme systems, by increase in temperature, exhaustion of the substrate or oxygen, and accumulation of carbon dioxide to inhibitory levels. Ching *(1961)* found that increased moisture accelerated respiration of perennial ryegrass and crimson clover seeds more than did increased temperature.

Preservatives for Reducing Respiration

Considerable research has been conducted to find a chemical that will inhibit the development of micro-organisms on seeds, grain, and oilseeds without adversely affecting their quality. Much of this work was done at the Department's Southern Regional Research Laboratory, New Orleans, La., by Altschul et al. *(1946)* and Lambou et al. *(1948)*. Preserving seed viability for propagation poses quite a different problem than controlling the growth of molds on grain and oilseeds to be used for milling and industrial purposes. Unfortunately no successful chemical or application has been found to control storage micro-organisms or respiration without damaging the seed embryo.

Conclusion

The most effective method of keeping the respiration of stored seed to a minimum is to keep the seed dry. It should be dried to a moisture content safe for storage and held at a relative humidity that will maintain a safe moisture content throughout storage. Most of the research on seed respiration has been on cereal seeds, oilseeds, and tree seeds. Nevertheless seeds of other plant groups or species have been the subject of interesting studies. For more information, see Miller *(1938)*, Crocker and Barton *(1953)*, and Milner and Geddes (*in* Anderson and Alcock, *1954*).

EFFECTS OF PESTS AND CHEMICALS ON SEED DETERIORATION IN STORAGE[9]

At harvest, seed lots contain much extraneous material, most of which is removed by cleaning. However, certain fungi, bacteria, viruses, and insects are not removed, and they cause or hasten seed deterioration. In addition, chemical treatments to control fungi, insects, and rodents may affect germination and longevity.

Seedborne viruses are rather uncommon. When present they usually remain dormant in stored seeds and infect the seedling subsequent to germination. Bacteria and actinomycetes, although capable of damaging stored seeds, usually remain quiescent because the relatively high moisture content necessary for their growth seldom occurs in commercial seed storage.

Storage Fungi

Saprophytic and parasitic seedborne fungi include *Alternaria, Chaetomium, Cladosporium, Fusarium, Helminthosporium,* and *Rhizopus* spp. They remain dormant during seed storage unless seed moisture content increases greatly, e.g., to over 14 percent in cereals. However, these fungi may prevent infection of grass seeds by storage fungi during the first part of the storage period (Kulik and Justice, *1967*). Certain molds (fungi), not usually present in or on seeds at harvest, can carry out their life cycle on stored seeds (fig. 6).

These fungi, termed "storage fungi" by Christensen and Kaufmann *(1965)* to differentiate them from the "field fungi," which are present in

[9]Prepared by M. M. Kulik, research plant pathologist, Agricultural Marketing Research Institute, Northeastern Region, Agricultural Research Service.

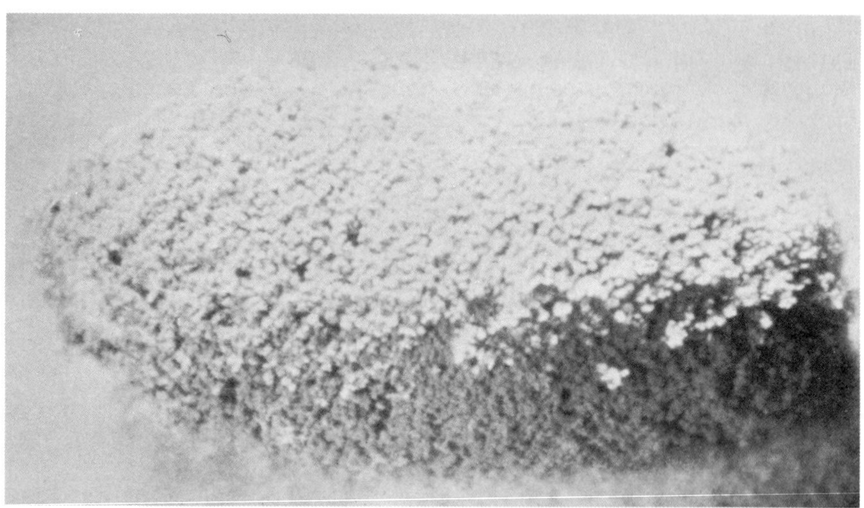

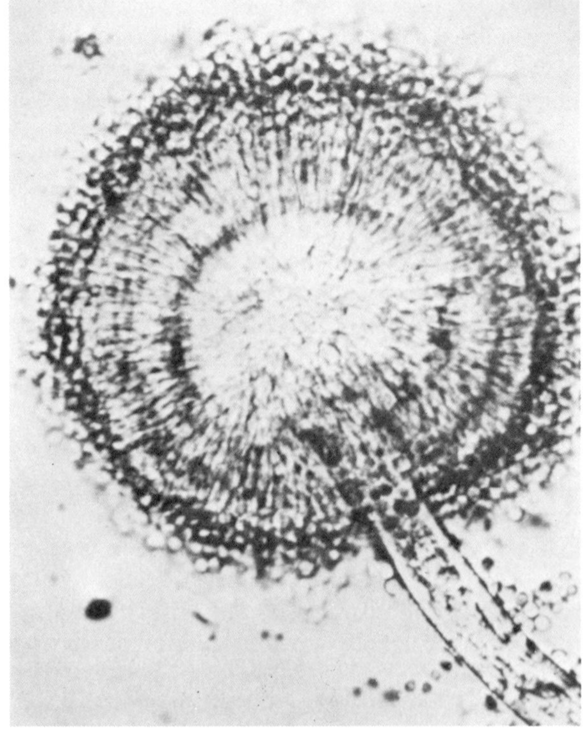

PN–5397, PN–5398

FIGURE 6.—*Above*, an enlarged cottonseed completely overgrown by the storage fungus *Aspergillus ochraceus*; *below*, its spore head greatly enlarged.

or on seeds prior to storage, can destroy stored seeds (table 16), because they can grow under limited moisture conditions where field fungi and other micro-organisms cannot grow. In fact, many of the storage fungi are actually osmophilic and grow best under relatively dry conditions (table 17). Some can invade seeds with moisture contents in equilibrium with an ambient relative humidity of 65 percent. Typical moisture contents are corn 13.8, oats 13.2, rye 14.3, and wheat 14.1 percent (Kreyger, *1954*). (All moisture contents in this section are on a wet weight basis.)

Storage fungi have been reported to invade and destroy cereal grains (Christensen and Kaufmann, *1969*), cocoa beans (Bunting, *1930*), cottonseeds (Arndt, *1946*), grass seeds (Kulik and Justice, *1967*), locust-beans and Scotch beans (Snow et al., *1944*), peas (Fields and King, *1962*), peanuts (Diener, *1958*), and soybeans (Milner and Geddes, *1946*). They can attack almost any kind of seed under favorable environmental conditions, since these molds can grow on most organic materials. Invasion of seeds by storage fungi may result in loss of viability, increase in free fatty acids, decrease in nonreducing sugars, development of musty odors, and discoloration. Deterioration can occur in a few days when seeds are stored under unfavorable conditions.

Storage fungi are principally *Aspergillus* and *Penicillium* spp., which commonly occur throughout the world. They are usually present in large numbers in the air and on surfaces in seed storage areas; they may be on a few seeds in a lot. They will invade and destroy seeds at 4°–45° C and 65- to 100-percent relative humidity. Their activity is largely determined by the physical condition, vitality, and moisture content of the seed and the ambient temperature and relative humidity of the storage area. Consequently, the fungal population reflects the kind and efficiency of the postharvesting handling, conditioning, and storage environment of a seed lot.

Much of our knowledge of storage fungi has come from studies of their activity in stored cereal seeds. Of the many *Aspergillus* spp., only a few have been reported to attack stored cereal seeds. *A. amstelodami*, *A. chevalieri*, *A. repens*, *A. restrictus*, and *A. ruber* grow in seeds with a moisture content of 13.2 to 15 percent. The major fungi found in cereal seeds above 15-percent moisture content are *A. candidus*, *A. flavus*, *A. ochraceus*, *A. tamarii*, and *A. versicolor* (Christensen, *1957*). *Penicillium* spp., though not as common as *Aspergillus* spp., are often found in cereal seeds, particularly in lots with a moisture content above 16 percent and stored at relatively low temperatures (Christensen and Kaufmann, *1965*).

A rapid growth of fungi in stored seeds can produce so-called hot spots caused by heating. Using seeds with 18-percent moisture content, Gilman and Barron *(1930)* found in laboratory tests that *A. flavus*, *A.*

TABLE 16.—*Frequency of* Aspergillus glaucus *in stored grass seed and germination after 2 to 12 months' storage under 16 temperature-relative humidity (RH) conditions*[1 2]

Storage temperature (°C)	35-percent RH		55-percent RH		75-percent RH		95-percent RH	
	Fungus frequency	Germination	Fungus frequency	Germination	Fungus frequency	Germination	Fungus frequency	Germination
	Percent	*Percent*	*Percent*	*Percent*	*Percent*	*Percent*	*Percent*	*Percent*
				2 MONTHS' STORAGE				
10	16	93	6	91	5	92	7	92
20	3	94	2	92	1	91	99	90
30	27	94	8	93	47	77	96	53
35	48	94	31	91	84	16	93	1
				4 MONTHS' STORAGE				
10	5	91	1	91	2	88	3	92
20	3	88	5	92	50	88	100	83
30	50	92	29	89	100	16	95	0
35	77	92	20	85	---	---	---	---
				8 MONTHS' STORAGE				
10	1	94	1	94	0	93	7	94
20	2	94	2	95	95	64	45	4
30	26	97	9	91	---	---	---	---
35	7	93	2	49	---	---	---	---
				12 MONTHS' STORAGE				
10	1	90	0	90	1	92	1	92
20	0	94	1	94	66	27	---	---
30	5	93	1	73	---	---	---	---
35	1	88	1	19	---	---	---	---

[1] Each value is average for *Festuca arundinacea*, *F. rubra*, *Lolium multiflorum*, and *Poa pratensis*; 100 seeds per sample cultured on 10-percent sodium chloride-malt agar after sodium hypochlorite treatment; 200 seeds per sample used to determine germination.

[2] Data from Kulik and Justice *(1967)*. (With permission of Microform Internatl. Mktg. Corp. exclusive copyright license of Pergamon Press Journal Back Files.)

TABLE 17.—*Colony diameter of mass isolates of 6* Aspergillus *species grown for 24 days at 22°–24° C on malt agar substrate*[1 2]

Aspergillus species and isolate (No.)	Colony diameter on malt agar with indicated sodium chloride (percent)				
	Control	5	10	15	20
	Mm	*Mm*	*Mm*	*Mm*	*Mm*
amstelodami					
1	31	82	85	86	74
2	56	79	84	88	72
3	49	81	86	89	74
Average	45	81	84	88	73
flavus					
1	86	84	76	30	6
2	85	83	58	0	0
3	90	90	90	40	0
Average	87	86	75	23	2
niger					
1	87	87	73	31	9
2	87	88	76	11	0
3	90	90	90	63	2
Average	88	88	80	35	4
ochraceus					
1	90	90	80	65	0
2	83	90	85	62	9
3	88	88	88	82	42
Average	87	89	84	70	17
parasiticus					
1	82	83	54	24	0
2	83	84	67	34	0
3	83	86	68	27	1
Average	83	84	63	28	0
terreus					
1	34	80	77	48	2
2	35	80	71	44	2
3	51	83	75	44	3
Average	40	81	74	45	2

[1] Malt extract 30 gm, agar 15 gm, water 1 l; 6 replications per isolate per treatment.
[2] Data from Kulik and Hanlin *(1968)*.

fumigatus, and *A. niger* raised the temperature of barley, oats, and wheat from 17° to 43° C. These results were confirmed and extended by Christensen and Gordon *(1948).* Mead et al. *(1942)* showed that the rise in temperature can be even higher if the seed contains a high proportion of broken kernels and cracked seedcoats. Similar hot spots can occur in bulk seed, which contains pockets of moist seeds, as the natural insulating properties of the surrounding dry seeds prevent rapid dissipation of the heat generated by respiration of the moist seeds and the associated fungi.

Pockets of moist seeds can arise in low moisture content seeds through roof leaks, insect activity, and moisture translocation when temperature gradients within the seed mass are allowed to occur. The fungi associated with hot spots in seeds stored on the farm were studied by Wallace and Sinha *(1962).* They observed that hot spots may occur at any location in a bin and that temperatures can increase to 53° C in winter. Field fungi such as *Alternaria* spp. were not common in the heated seeds, although they were present in many of the viable seeds. The heated seeds were infected chiefly by storage fungi, mainly *Penicillium* and *Aspergillus* spp., especially *A. flavus, A. fumigatus,* and *A. versicolor. Absidia* spp. were also frequently observed. Christensen *(1957)* stated that "heating is likely to be the final and violent effect of mold invasion of the seed, not an indication of beginning deterioration."

Sinha and Wallace *(1965)* investigated the succession of fungi and actinomycetes populations associated with progressive stages in the heating process of an artificially caused fungus hot spot in stored wheat. The growth of *Penicillium cyclopium* and *P. funiculosum* in a moist pocket of grain stored at −5° to 8° C for 4 months began the heating process. A maximum temperature of 64° was reached in this hot spot. Since these two fungi cannot live above 55°, bacterial and biochemical activity was deemed responsible for the increase to 64°. The following ecological succession of micro-organisms, which often overlapped, was observed: *Penicillium cyclopium, P. funiculosum, Aspergillus flavus, A. versicolor, Absidia* spp., and *Streptomyces* spp.

Several investigators have studied the mycoflora of freshly harvested and stored peanuts. Hanlin *(1966)* observed the internally borne mycoflora of Georgia peanuts. *Fusarium* spp. were found in 20.7 percent, *Penicillium* spp. in 18.5 percent, and *Aspergillus* spp. in 7.5 percent of the seed at harvest. Jackson *(1965)* used dilution methods to investigate the mycoflora of soil that adhered to dry peanut pods. Relatively small numbers of *Aspergillus flavus, A. niger, A. terreus, Rhizopus* spp., and *Sclerotium bataticola* were found. *Penicillium citrinum, P. funiculosum, P. rubrum,* and *Fusarium* spp. were found in large numbers. *Aspergillus flavus, A. niger, Rhizopus* spp., and *S. bataticola* extensively penetrated pods and kernels when dry; the infested pods were hydrated for 6 days at 26°, 32°, or 38° C. As the temperature increased, infection by *A. flavus* and *A. niger* increased.

Garen *(1966)* studied the endogeocarpic floras of Virginia peanuts and reported that "*Trichoderma viride* seems a dominant and *Penicillium* spp. seem sub-dominants in the climax endogeocarpic community of sound and rotting peanut pods; and *Aspergillus flavus* and *A. niger*, which have a potential for causing trouble, are quantitatively minor but persistent species in this flora."

Diener *(1960)* investigated Georgia farmers' stock peanuts (uncleaned and unshelled) that had been stored for 8 to 56 months. He found that the predominant fungi consisted of certain species of the *Aspergillus glaucus* group (*amstelodami, chevalieri, repens,* and *ruber*), *A. restrictus, A. tamarii, Cladosporium* spp., *Penicillium citrinum, Torula sacchari,* and members of the Mucorales. Large numbers of fungi were directly associated with kernel moistures of 12.5 percent or greater at the time the seeds were stored. Kernel moisture content decreased to about 7 percent, which is considered safe, in 3 to 4 weeks after the bins were filled, regardless of the initial moisture content of the kernels. During the 5-year storage period the moisture content of no sample ever exceeded 6.8 percent. There was no relationship between percentage of initial kernel damage, final damage, or increased damage during storage.

Control of Fungal Deterioration

Deterioration of stored seeds by fungi is controlled principally by drying the seeds to a safe moisture content prior to storage in a dry place. Most storage fungi cannot invade seeds that are in moisture equilibrium with 65-percent relative humidity or lower. However, some fungi are xerophytic and can invade seeds that contain only slightly too much moisture.

Because the moisture content of a seed lot represents the average of many seeds, some individual seeds may possibly contain an unsafe amount of moisture even though the moisture content of the lot is at a safe level. Seeds containing too much moisture are susceptible to invasion by storage fungi, which may then spread throughout the lot. Therefore it is extremely important to accurately measure seed moisture content. For best results, seeds should be dried as soon as possible after harvest, then thoroughly cleaned prior to storage in an environment where they will not absorb moisture.

Fumigation to control insects ordinarily has no detectable effect on storage fungi (Christensen and Hodson, *1960*).

For the effects of micro-organisms on stored cereal seeds, see Semeniuk *(1954)*, Christensen and Kaufmann *(1969, 1974)*, and Wallace *(1973)*.

Do not assume that the effects of storage fungi on seeds of other field crops, flowers, or vegetables will necessarily be the same as those on stored cereal seeds. For example, Kulik *(1973)* found that seeds of cabbage, cucumber, pepper, radish, and turnip inoculated with spores of either of two common storage fungi and stored at 85-percent relative humidity and 22°–25° C for 30 days remained largely free of invasion. Qasem and Christensen *(1958)* found that 84 percent of the corn seeds they inoculated with a storage fungus and stored at 85-percent relative humidity and 20° were invaded by the fungus after 1 month. Also, Tuite and Christensen *(1957)* found that 100 percent of the wheat seeds they inoculated with a storage fungus and stored at 80-percent relative humidity and 25° were infested after 1 month.

Insects

As with storage fungi, much of our knowledge of insects that attack stored seeds comes from studies of stored grain and grain products. According to Henderson and Christensen *(1961)*, stored seeds are attacked primarily by those insects that destroy stored grain and grain products. Of the several hundred insect species associated with stored grains and seeds, only about 50 cause problems. Of that 50 only about a dozen cause serious damage, including the rice weevil, granary weevil, lesser grain borer, Angoumois grain moth, cadelle, sawtoothed grain beetle, khapra beetle, and flat grain beetle. The weevils puncture the seedcoat and destroy the endosperm, but other stored-product insects attack the embryo. In either case the germination potential is reduced or totally destroyed.

Stored-product insects that fly migrate to previously noninfested seed storage areas. Seeds may also become infested from contaminated storage bins or sacks or in the field prior to harvest. One species of chalcid lays its eggs individually in seeds of alfalfa, certain clovers, and trefoils. The larva gradually consumes the contents of the seed and becomes an adult wasp. It then chews its way out of the seed, mates, and the female lays her eggs in host seeds, whether on the plant or in a seed warehouse.

Control measures recommended by Lieberman et al. *(1961)* are designed to destroy as many as possible of the larva-infested seeds: "One should clean all seed carefully and destroy or use the cleanings, prevent seed from forming on volunteer plants, clear the field after harvest, work all chaff into the soil, and, if necessary, provide moisture to encourage the growth of fungi present in the soil that will kill the overwintering larvae in the seeds."

Additional insect control measures include thorough cleaning and fumigation of all seed-handling equipment, seed containers, and storage areas, and when required an insecticide application.

For some types of insects, fumigating the seeds before or during storage is also essential to good control. Seed fumigation must be very carefully done, as some fumigants under certain conditions are very harmful to seed germination. Methyl bromide is especially harmful to seed germination when improperly applied.

Cotton *(1954, 1960)* reported on the role of insects in deterioration of stored seeds.

Rodents

Rats, mice, and squirrels destroy thousands of pounds of seed each year. Much of the loss is not what these rodents eat, but what they scatter and mix. They have been reported to destroy as much as 200 million bushels of grain each year in the United States (Cotton, *1960*). Even greater losses occur in other parts of the world. The best control for rodents is to keep them out of seed storage areas. This can be accomplished by rodentproof construction and by eliminating such entryways as cracks and holes in walls and floors and unscreened ventilators. Cleanliness inside and outside storage areas also helps to keep rodents away.

Rodent infestations can be eliminated by using traps, poison baits, or fumigation. Generally toxic gases are handled by an expert. Spencer *(1954)* and Cotton *(1960)* published detailed accounts of rodents in grain and their control.

Fungicides and Fumigants

Seed treatments have long been recognized as injuring seeds in storage. Clayton *(1931)* reported that the recommended treatment for many kinds of seeds with mercuric chloride, liquid organic mercurials, or hot water greatly reduced seed longevity. However, seeds treated with fungicide dusts usually suffered no injury from the treatment. Crosier *(1938)* found abnormal germination in wheat seeds that had been treated with Ceresan and held in cool, dry storage for 1 year prior to being tested for viability. Corn seed of 12-percent moisture or less was treated by Koehler *(1938)* with several mercurials and stored up to 1 year. When treated seed was stored near the roof of an open shed, it was definitely injured every time. In general, in this study, freshly treated seeds performed better by a small margin.

Miles *(1939)* showed that storage of cottonseed treated with Ceresan or New Improved Ceresan in a dry laboratory up to 17 months caused no injury to the seed. In another study, Miles *(1941)* stored mechanically delinted cottonseed treated with New Improved Ceresan in 100-pound bags (storage conditions not given). Germination tests showed no injurious effects after 17 months in cotton bags.

Seeds of alfalfa, ladino clover, red clover, and sudangrass were treated with New Improved Ceresan, Arasan, Spergon, Semesan, or yellow cuprocide by Kreitlow and Garber *(1946)* and were stored in closed or open containers at 10° or 25° C for 30 months. At 25°, only the germination of sudangrass treated with New Improved Ceresan was greatly reduced. At 10°, germination of treated seeds and controls stored in closed containers was much better than that of seeds stored in open containers. Jute seeds treated with a mercuric acetate or ethyl mercuric chloride dust were stored by Ghosh et al. *(1951)* for more than 14 months in airtight, moisture-free, nonporous containers without any reduction in viability.

Koehler and Bever *(1954)* reported reduced germination of oats and wheat caused by volatile mercury compounds. The reduction increased as dosage, length of storage, and moisture content of the seed increased. Compounds differed in the amount of injury they caused. Reduced viability was not always related to volatility of the compound. In another study, wheat seeds of two cultivars of about 10-percent moisture content were treated by Koehler and Bever *(1956)* with 11 organic mercury fungicides and stored for 10 months in stoppered bottles at 4.5° or 21° C. Germination tests in sand showed that most of the treatments caused significantly lower stands than in the nontreated controls. Also, there were striking differences among the compounds in the amount of injury they caused. In general, however, they found that seed storage at low temperature resulted in less seed injury.

Nakamura et al. *(1972)* treated seeds of beet, cabbage, carrot, eggplant, pea, pepper, and tomato with four organic mercury and four organic sulfur fungicide dusts. The seeds were then stored for up to 3 years. Organic mercury compounds greatly reduced the viability of eggplant seeds and caused some injury to seeds of beet and cabbage. Seeds of carrot, pea, pepper, and tomato were not injured by these compounds. Seeds of these vegetables treated with organic sulfur dusts and stored for 1 year did not undergo reduced germination. Properly applied fungicides usually do not impair germination during favorable storage conditions.

Young *(1929)* investigated the effect of several fumigants on seed viability for controlling rice weevils. He found that seeds of alfalfa, barley, beans, buckwheat, clover, corn, cowpeas, lima beans, oats, rye, sunflower, timothy, and wheat could be treated with up to twice the minimum lethal concentration of ethylene dichloride, isopropyl formate, tertiary butyl chloride, and trichloroethylene without seriously reducing their viability. Ethylene oxide and methyl chloroacetate seriously reduced germination.

Cobb *(1958)* fumigated seeds of 14 field crops and 15 vegetable crops at 3 moisture contents with methyl bromide. Germination tests were made immediately after fumigation and after storage for up to 90 days. Some seeds were stored for up to 2 years. He found that injury occurred while seeds were in contact with the methyl bromide, that fumigant injury was not reversible, that seeds uninjured by fumigation did not exhibit subsequent injury, and that some seeds injured by fumigation declined in viability more than the nontreated control seeds. Based on his study, Cobb recommended that "the fumigation of dry seed at low temperatures and gas concentrations offers the greatest opportunity for the safe use of methyl bromide." He also recommended "that for fumigation when seed viability is important, test fumigation and germination should be done before large-scale fumigation is attempted with methyl bromide."

King et al. *(1960)* evaluated hydrogen cyanide, methyl bromide, and carbon tetrachloride alone and in mixtures with several other fumigants for their effect on germination of barley, corn, cotton, oats, rice, sorghum, and wheat at three seed moisture levels and three treatment temperatures immediately after fumigation and after storage in polyethylene bags for 12 months at room temperature. They found that either a high seed moisture content or a high temperature during fumigation caused a reduction in viability. A high seed moisture content combined with a high temperature usually resulted in extensive injury. Fumigated seeds of all seven crops produced less vigorous seedlings than did the nonfumigated controls, although in some cases differences in germination percentages between the fumigated and nonfumigated seeds were minor. The rate of germination of stored seeds was decreased by fumigation. The authors recommended that "germination tests should be performed immediately before sale if fumigated seed have been in storage for a prolonged period."

Potgieter and de Beer *(1972)* treated white corn, yellow corn, and grain sorghum with methyl bromide gas or phosphine gas, then planted the seed in a sandy loam soil after about 1⅜ inches of rainfall. Methyl bromide fumigation was rather injurious and repeated fumigation with phosphine decreased field performance.

CHANGES ASSOCIATED WITH SEED DETERIORATION

Biochemical Changes

Seed storage begins immediately after maturity regardless of where or how seeds are held. During seed development anabolic processes predominate and bring about a gradual increase in dry matter, including development of an embryo and food reserves. Following maturation biochemical changes continue and eventually catabolic processes predominate and deterioration becomes apparent. Catabolic changes occur more slowly under low temperature and low relative humidity than under high temperature and high relative humidity.

Although much has been written about chemical changes associated with seed deterioration (Crocker and Barton, *1953*; Zeleny *in* Anderson and Alcock, *1954*; Abdul-Baki and Anderson *in* Kozlowski, *1972*; Roberts, *1972*), our knowledge of biochemical deterioration of seeds is still inconclusive. Attempts have been made to correlate such factors as depleted food reserves, increased enzyme activity, increased fat acidity, membrane permeability, and similar changes with deterioration. Biochemical changes in the aging process take place in both the embryo and endosperm. By embryo transplant technique Kikuchi *(1954)* and Floris *(1970)* found that embryos from new seeds grew slower when transplanted to old ("decrepit") endosperm than when transplanted to the endosperm of new seeds. Kikuchi *(1954)* reported that the embryo deteriorated more rapidly than the endosperm, and he theorized this was because of fundamental differences in their nature. Floris *(1970)* concluded that aging is a progressive process accompanied by accumulation of toxic metabolites, which progressively depress germination and growth of seedlings with increased age. Obviously important biochemical changes take place in the embryo and endosperm during storage, but the significance of these changes in relation to deterioration is not fully understood.

Villers (*in* Heydecker, *1973*) proposed that deterioration in dry stored seed results from the lack of operable systems to repair and replace organelles, whereas in moist stored seeds repair systems are fully operable. He pointed out that the full consequences of damage accumulating during storage are not evident until imbibition.

Germination and Vigor

From prehistoric times man has known that the germination capacity of seeds declined with age (James, *1967*). Even today most studies of physiological and biochemical changes in seeds, oilseeds, and grain include plans for determining the percent germination as a criterion of deterioration or change.

Vigor is considered here in relation to the seed's strength or power of germination, its ability to send out a strong root and shoot even under conditions of stress, and its freedom from attack by micro-organisms.

Loss in vigor can be thought of as an intermediate stage in the life of the seed, occurring between the onset and termination of decline (death). Decline in vigor is extremely difficult to measure. No generally accepted method has yet been found to measure vigor, but the most useful available methods are based on measurements relating to germination capacity. For example, Abdalla and Roberts *(1969b)*, working on seed deterioration of barley, broadbeans, and peas, reported that the percentage of seed viability is an excellent indicator of growth potential of the surviving seeds, irrespective of the factors responsible for the loss of viability. Gill and Delouche *(1973)* found rate of germination and seedling growth of corn to be the most consistent and sensitive measures of the progress of deterioration. A similar conclusion was drawn by Christidis *(1954)*, who found that the rate of germination of stored cottonseed increased for the first 5 years of storage and then gradually decreased throughout the following 6 years.

As the catabolic changes continue with increasing age, the ability of the seed to germinate is reduced. The curve in figure 2 shows that measurable decline in viability or germination capacity does not begin immediately after maturity. Under favorable storage conditions the initiation of decline may be a few months to many years depending on storage conditions, kind of seed, and previous storage conditions. When making germination tests on stored seed, one of the first indications of deterioration is reduced vigor, which is shown by reduced rate of germination and production of weak or watery seedlings and seedlings with stunted radicles. Seeds with reduced vigor would appear to produce low yields compared with fresh vigorous seeds; however, this is not always true.

Yield

Old seed can be vigorous and alternatively new seed can be deteriorated depending on the storage conditions to which they have been subjected. Therefore seed of a given chronological age may or may not be deteriorated. Some of the pertinent literature does not make this distinction. If a maximum yield of a given crop is to be obtained, a minimum number of plants per acre (or hectare) is necessary. If the rate of planting low germinating seed is increased to satisfy this minimum requirement, yield is not decreased compared with the yield from new, vigorous seed. The problem is whether plants from seeds of low vigor can compete in the field with plants from vigorous seeds.

Abdalla and Roberts *(1969a)* planted seeds of barley, broadbeans, and peas in a culture medium and in the field. Plants from some seeds associated with decreased viability exhibited retarded growth of roots and shoots and increased variability in growth during early life. The early inhibition of growth rate did not persist, and there was evidence that under normal cultural conditions the low growth rate during early plant development might be compensated for in later growth stages. They concluded that unless the germination of deteriorated seed of barley, broadbeans, and peas is less than 50 percent, the final yields are not significantly affected.

Christidis *(1954)* found that plants from cottonseed up to 10 years old yielded about the same weight of seed cotton as plants from fresh seed. Others who have reached similar conclusions from work with seeds of different crop species are Rodrigo *(1939)* with mung beans, Barton and Garman *(1946)* with aster, lettuce, pepper, and tomato, Fulutowicz and Bejnar *(1954)* with sugar beet, and Ghosh and Basak *(1958)* with jute. Barton and Garman *(1946)* found yields of crops from seeds of the four species, which had been stored at −5° C up to 13 years, were equal to those of fresh seeds. Thirteen-year-old lettuce seed produced larger heads than did fresh seeds. Likewise plants from the 11-year-old mung bean seed used by Rodrigo *(1939)* produced significantly higher yields of pods and beans than plants grown from fresh seed.

Not all the evidence is in favor of normal yields of crops produced from deteriorating seed. Newhall and Hoff *(1960)* stored onion seeds for 22 years over calcium chloride in a closed desiccator. Field tests revealed a 30-percent loss in yield by the 22-year-old seed compared with yield of fresh seed. Bulbing vigor was reduced by 15 percent, although this was hardly detectable by casual observation.

Chirkovskiy *(1953)* reported that longtime storage of tobacco seed resulted in loss of germination and lower yields of leaves per plant compared with yields of plants from fresh seeds. He *(1956)* indicated that aging of seeds had a depressing effect on plant development, especially during early growth and that unfavorable storage conditions increased the depressing effect.

Corn that germinated only 38 percent in the field yielded only 50 percent as much grain as seed that germinated 80 percent in the field (Heit and Munn, *1947*). It is not clear whether the reduced yield was due to reduced field stand, weakened vigor, or both. They also claimed reduced yields from aging seeds of cucumber and soybean. Presumably they credited the reduced yields to low seed vigor, as seeds of both crops produced seedlings with stubby roots insufficient to properly anchor the plant or to provide an adequate supply of soil nutrients.

Because information is limited on plant yield from seeds of low or weakened vigor, future research should segregate the principal factors so the contribution of each can be ascertained. Among these factors are use of uniform genetic stocks, isolation of low vigor effects, and effects of competition versus noncompetition with other plants under normal field conditions.

Cytological Changes

One of the changes associated with seed aging is aberration of chromosomes, sometimes referred to as mutagenic effects. de Vries is credited with discovering mutations in aging seeds in 1901 (Kostoff,

1935). Some of the chromosome alterations in aged seed include fragmentation, bridges, fusion, ring formation of chromosomes, and variation in nuclei size. These changes have been found in a relatively large number of crop and native plant species, including *Antirrhinum*, *Crepis*, *Datura*, *Nicotiana*, barley, broadbean, corn, onion, pea, rye, sugar beet, and wheat.

According to James *(1967)*, there is no evidence that these cytological aberrations develop in seeds stored under favorable conditions. The seeds used in most experimental work so far have been subjected to high storage temperatures or humidities or both. Possibly James' statement can be questioned depending on what is considered favorable or unfavorable storage conditions. For example, Blakeslee *(1954)* found mutations in *Datura* seeds stored at room temperature. He did not indicate that either the temperature or the relative humidity was unfavorable during storage. He also found a very low frequency of mutations in *Datura* seeds that had been buried in the soil for 39 years.

HOW SEEDS ARE DRIED

Characteristics of Water

Because drying involves removing water from the seed, let us first consider some of the characteristics of water. In nature, water exists simultaneously as a solid, liquid, and gas. Water freezes at 0° and boils at 100° C. It contracts as it cools to near the freezing point but expands on freezing. The quantity of solid or liquid water is expressed by weight or volume, whereas the quantity of gaseous water is given as percent humidity, either relative or absolute. The amount of water in a seed is its moisture content, which is measured on a wet or dry weight basis with respect to the seed weight. When seed moisture content is given on the wet weight basis, the amount of water in the seed is a percentage of the seed weight before the water is removed. When seed moisture content is expressed on the dry weight basis, it is a percentage of the seed weight after the water is removed.

During development, maturation, and ripening, the water content of seeds gradually decreases until harvested seeds finally dry to the point where there is no further reduction in moisture because their water content reaches equilibrium with the ambient relative humidity. Any further change in seed moisture content will be the result of a change in either the relative humidity or the temperature of the air or both. Because of chemical differences in their composition, not all kinds of seeds will equilibrate at the same moisture content at a given relative humidity. In fact, not all cultivars of a given kind of seed nor even all lots of a cultivar have exactly the same equilibrium moisture content for any given relative humidity.

Water in seeds is present in liquid form in cell walls and cell contents and in gaseous form in intercellular spaces. Since water is removed from seeds by evaporation, it is removed only in its gaseous form.

Principles of Drying

Seed drying requires the transfer of heat, because a seed can be dried only by evaporating moisture from its surface, and the heat content of moisture vapor is greater than that of liquid water. Heat transfer may be accomplished by contact, convection, or radiation. Most seed-drying equipment transfers heat by convection. Drying of a seed requires that evaporation of moisture from its surface be accompanied by the transfer of moisture from its interior to its surface. When water evaporates from the seed surface into the atmosphere, a moisture gradient is set up inside the seed that causes internal moisture to move toward the surface. If evaporation from the seed's surface occurs too rapidly, the extreme moisture stress that develops can, and usually does, damage the embryo and cause loss of viability. It is therefore essential that seed drying be carefully controlled to prevent stress damage. It is also important that the vapor pressure of the surrounding atmosphere not be allowed to increase. If the vapor pressure of the atmosphere becomes greater than that of the seed surface, the seed will gain rather than lose moisture. Additional information on principles of drying seeds can be found in Anderson and Alcock *(1954)*, Hall *(1957)*, Brandenburg et al. *(1961)*, Kreyger *(1963a)*, Harrington and Douglas *(1970)*, and Philpot *(1976)*.

Methods of Drying Seeds

Methods of drying seeds include natural drying, sun drying, unheated, heated, and dehumidified air-drying, drying in storage and before storage, drying with desiccants, vacuum drying, and freeze drying. These methods may be used alone or in various combinations.

Natural Drying

Natural drying occurs in the field during ripening and in the bin after harvest. It is the process by which seeds lose water naturally without help from man. The extent of natural drying is regulated by such factors as air temperature, relative humidity, and wind velocity. Shatter resistance, harvesting procedures, and workload of the producer also affect the amount of natural drying permitted before harvest. A hot, dry wind prior to full maturity can seriously damage a seed crop by causing too rapid drying, but a moderate temperature and relative humidity can produce top-quality seed. Some but not all seed crops are harvested or partially harvested before the seeds dry to moisture equilibrium with the atmosphere. Such seeds must receive additional drying. Sometimes the plants are cut and placed in windrows and this speeds up natural drying.

Sun Drying

Where sun drying is possible, seeds may be spread in screen-bottom trays placed on low racks in the sun or on flat-roofed houses, or they may be spread on the ground or on concrete. Seeds dried in the sun are stirred from time to time to facilitate rapid, uniform drying. Sun drying is used for such seeds as cantaloup, cucumber, pumpkin, tomato, watermelon, and others that are threshed wet. Because these seeds do not separate readily from the surrounding pulp, fermentation, chemical treatment, or mechanical action is used to help free them. Once freed, they are washed thoroughly and placed in screen-bottom trays to be cured in the sun. Although these and other kinds of seed are frequently dried naturally, they may also be dried artificially. Most seeds harvested at a moisture content too high for safe storage are dried artificially. Solar heat can be used in the bin drying process (fig. 7).

Artificial Drying

Three basic types of seed dryers are layer, batch, and continuous flow. There are sound engineering and economic reasons why each type is better suited for certain kinds of drying than the others. No one type of dryer is best suited for all drying needs. The user must choose the best one for his specific needs. One should not attempt to select a seed dryer without first obtaining sufficient information on the advantages, disadvantages, and limitations of all available dryer systems. One must also have sufficient information on the specific drying requirements of the kind or kinds of seeds to be dried. Table 18 presents data for several crops. Although not complete, the data provide a reference for related kinds of seeds.

Layer Drying.—Layer drying is usually in-storage drying, which may be done with or without supplemental heat. For layer drying, the storage area is equipped with a network of air ducts attached to a large fan. The air ducts (fig. 8) should be placed on the floor of flat storage structures so they provide reasonably uniform airflow, cooling, and drying in the entire mass of seeds. The manifold should be as close to the wall as possible, and in round buildings it should be opposite the door so as not to interfere with unloading. The method of unloading must always be considered when locating the aeration tubing. Other factors such as depth of seeds, accumulation of fine particles, and uneven loading have a direct bearing on operation efficiency of the drying system. The size and speed of the fan are determined by the size of the storage area, the kind and depth of seed to be dried, the

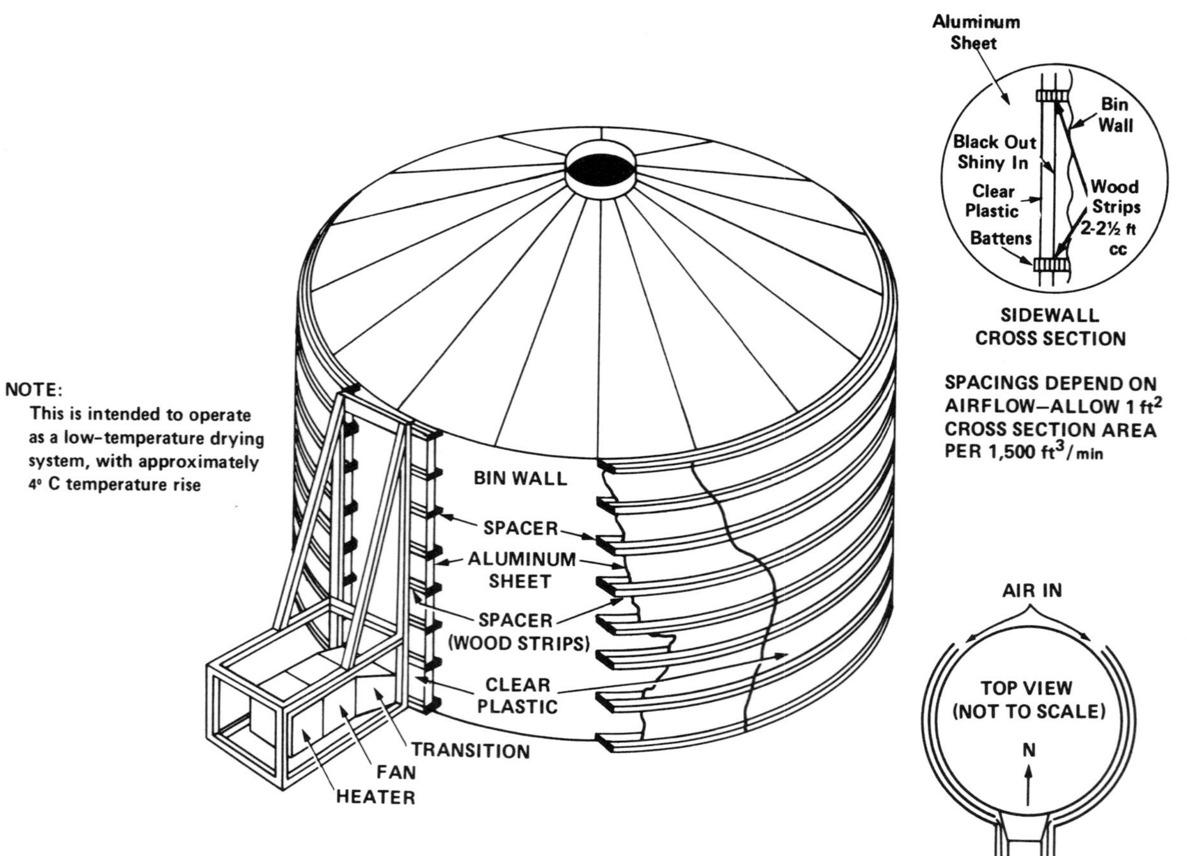

FIGURE 7.—Solar heat dryer for seeds and grains. (Courtesy of W. W. Peterson, S. Dak. State Univ.)

TABLE 18.—*Static pressures and estimated quantities of various seeds and grains that can be dried per batch per fan horsepower for different airflows and seed depths*[1]

Seed and airflow per bushel (ft³/min)	Seed depth	Static pressure[2]	Maximum quantity that can be dried per fan horsepower[3]
	Ft	*Water gage in*	*Bu*
Alfalfa:			
5	1	0.66	960
	2	1.93	330
10	1	1.09	290
	2	3.75	80
Blue lupine:			
5	2	.32	1,990
	3	.41	1,550
	4	.56	1,140
	6	1.15	550
	8	2.25	275
10	1	.29	1,100
	2	.42	760
	3	.74	430
	4	1.10	270
	6	2.83	110
	8	3.65	85
20	1	.35	470
	2	.75	210
	3	1.60	100
	4	3.05	50
Crimson clover:			
5	1	.52	1,225
	2	1.39	460
	3	2.95	220
10	1	.82	390
	2	2.65	120
Fescue, Ky. 31:			
5	1	.37	1,720
	2	.77	830
	3	1.51	420
	4	2.45	260
10	1	.51	625
	2	1.35	240
	3	3.04	100
20	1	.80	200
	2	3.05	50
Kobe lespedeza:			
5	1	.33	1,930
	2	.59	1,080
	3	1.06	600

See footnotes at end of table.

Table 18.—*Static pressures and estimated quantities of various seeds and grains that can be dried per batch per fan horsepower for different airflows and seed depths*[1]—Continued

Seed and airflow per bushel (ft³/min)	Seed depth	Static pressure[2]	Maximum quantity that can be dried per fan horsepower[3]
	Ft	*Water gage in*	*Bu*
Kobe lespedeza—Continued			
	4	1.81	350
	6	4.03	160
10	1	.42	760
	2	1.00	320
	3	2.14	150
	4	3.93	80
20	1	.62	260
	2	2.09	75
Lespedeza sericea:			
5	1	.66	960
	2	1.93	330
10	1	1.09	290
	2	3.75	80
20	1	2.00	80
Oats:			
5	2	.47	1,350
	3	.82	780
	4	1.29	490
	6	3.01	210
10	1	.36	880
	2	.77	410
	3	1.63	200
	4	3.05	100
20	1	.51	310
	2	1.65	100
Rescuegrass:			
5	2	.33	1,930
	3	.46	1,380
	4	.65	980
	6	1.21	525
	8	2.25	280
10	1	.29	1,100
	2	.44	720
	3	.73	430
	4	1.25	250
	6	3.01	100
20	1	.34	470
	2	.75	200
	3	1.63	100
	4	3.05	50

See footnotes at end of table.

TABLE 18.—*Static pressures and estimated quantities of various seeds and grains that can be dried per batch per fan horsepower for different airflows and seed depths*[1]—Continued

Seed and airflow per bushel (ft^3/min)	Seed depth	Static pressure[2]	Maximum quantity that can be dried per fan horsepower[3]
	Ft	*Water gage in*	*Bu*
Soybean:			
5	2	.33	1,930
	3	.43	1,480
	4	.62	1,025
	6	1.25	510
	8	2.33	270
10	1	.29	1,100
	2	.44	720
	3	.70	450
	4	1.29	250
	6	3.19	100
20	1	.35	470
	2	.77	200
	3	1.72	90
	4	3.25	50
Wheat:			
5	1	.33	1,930
	2	.59	1,080
	3	1.06	600
	4	1.85	340
10	1	.42	760
	2	1.05	300
	3	2.11	150
20	1	.65	240
	2	2.25	70

[1] Data from Wheeler and Hill *(1957)*.

[2] Includes 0.25-in allowance for loss from duct friction.

[3] Airflow (ft^3/min) per horsepower based on formula

$$ft^3 \text{ min} = \frac{\text{static efficiency}}{0.000157 \times \text{static pressure}},$$ with static efficiency assumed here at 50 percent.

temperature of the air, and the amount of air movement desired. Each layer of seed is partially dried before the next is added. The rate at which the dryer can be filled depends on the amount of moisture to be removed and how rapidly it is removed. The entire depth of seed is finally dried in place. Because the dried seeds remain in the dryer, this type of dryer is not satisfactory for large operations or for those that require a variety of drying conditions.

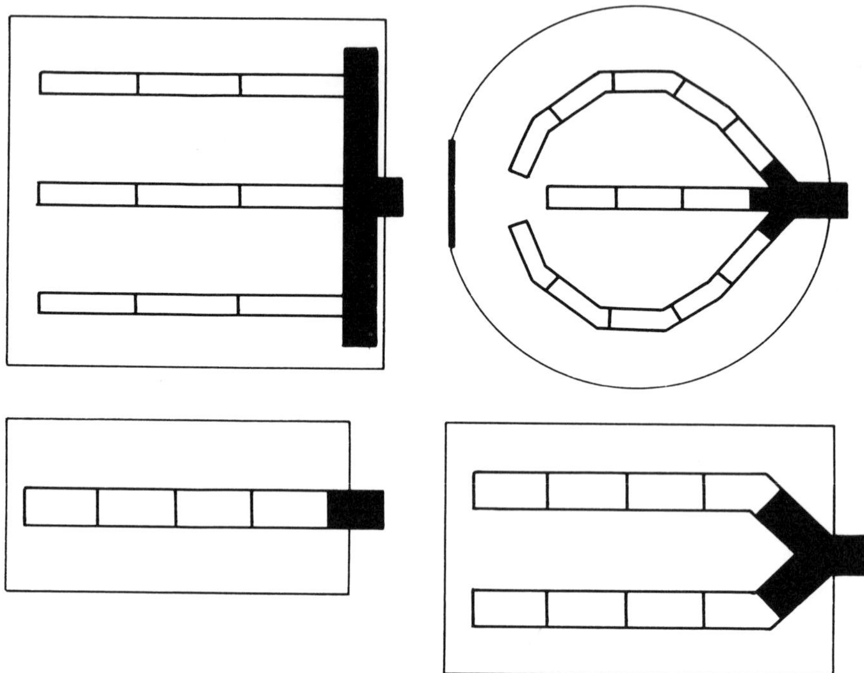

FIGURE 8.—Diagram of air duct arrangement for square, round, and rectangular in-storage seed and grain dryers.

Harvesting for storage in layer dryers can be accelerated by using several bins, supplemental heat, or both (fig. 9). However, the amount of heat used must not overdry the lower layers of seed before the top layers are dried. One cannot simply turn up the heat to dry faster because this may seriously overdry the lower layers. According to McKenzie *(1966)*, layer drying, contrary to popular belief, requires superior management. Layer drying is suitable for on-the-farm drying of corn, soybean, and cereal grain seeds.

Batch Dryers.—Batch dryers may be either portable or stationary. Portable batch dryers may or may not have a burner attached. Some batch dryers are specially designed wagons with porous beds that permit heated or unheated air to be forced upward through the seeds, whereas others are simply drying chambers. The wagon-type dryer can be taken to the field for filling, whereas the other types cannot. In any case, the batch-type dryer can dry only a fixed amount of seed at a time. After drying, the seed has to be cooled before it can be transferred to storage. Because of the time involved in drying and cooling, it is necessary to have several drying chambers for efficient harvesting and drying of seed crops. Since stationary batch dryers usually have a

FIGURE 9.—In-bin layer drying with supplemental heat. Bins are filled in rotation, one layer at a time. As each layer partially dries, successive layers are added until storage is full and the grain is dried in place. (Courtesy of Bruce McKenzie, Purdue Univ.)

greater capacity than portable dryers, they are preferable for commercial installations.

Stationary batch dryers may be a bin, column, or rotary drum. In bin dryers the seeds are spread over a perforated floor, and in a column dryer they stand in a vertical column. Depth of seeds over the floor or in the column varies with the type of seed being dried. In the rotary dryers a revolving, perforated drum slowly tumbles the seeds as heated air passes through them.

Batch-type dryer installations vary from a single drying chamber to many such chambers. Batch-type drying plants generally utilize one of five drying systems: (1) Double or two pass, (2) single-pass reversing, (3) single pass, (4) suction, or (5) tunnel. Batch dryers are suitable for both on and off the farm drying of many kinds of seeds, including cereal grains, corn, millet, sorghum, and soybean.

In the double- or two-pass drying system (fig. 10), the heated drying air is first directed through a bin containing nearly dry seeds, where the air picks up a small amount of moisture and loses a little of its heat. The air is then transferred and exhausted through a bin of high moisture content seeds, which need warming up. The double pass makes a more

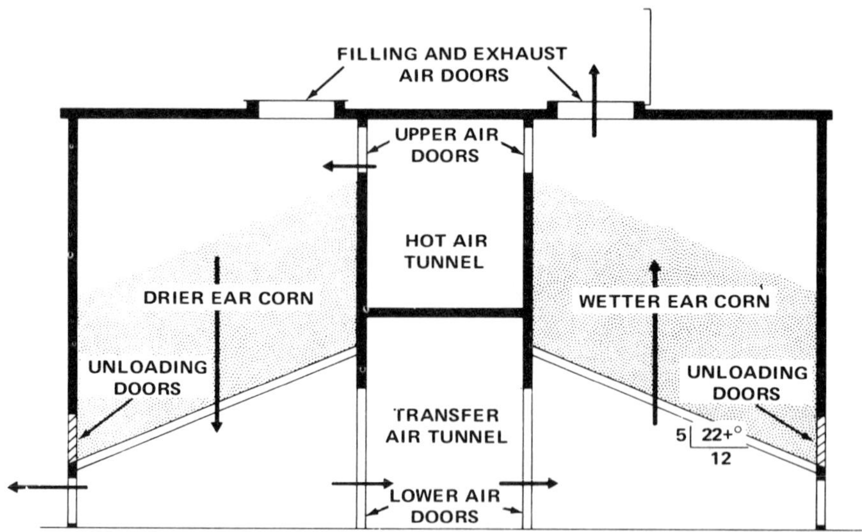

FIGURE 10.—Double- or two-pass drying system. (Courtesy of Corn States Hybrid Serv., Des Moines, Iowa.)

complicated drying system; however, it adds about 25 percent to the drying capacity of a drying plant and reduces fuel costs substantially, although power costs are increased some. The two-pass system provides added protection to the viability of the seeds by using cooler air (27°– 32° C) on the higher moisture seeds. Only the relatively dry seeds are subjected to a possible 43°.

For the single-pass reversing drying system (fig. 11), the seed drying plant is designed to allow a single pass of air with provisions for reversing the direction of airflow through the various bins to facilitate uniform drying. In the single-pass system the air goes through the seed and is exhausted. During the early stages of drying, a great deal of heat is transferred from the air to the seeds and considerable water is taken up by the air; however, toward the end of the drying cycle, little heat or moisture is transferred, resulting in considerable drying potential being wasted. This method has approximately 80 percent of the drying efficiency of the two-pass system.

The single-pass system (fig. 11) has no provision for reversing airflow; consequently, it has only about 75 percent of the efficiency of the two-pass system. With this system the seeds on the bottom of the bin will be overdried and those on the top underdried. A uniform moisture content is obtained by blending after the seeds are removed from the dryer.

In the suction drying system (fig. 12) the blower is placed at the exhaust side of the drying bins. This system can be used with either the

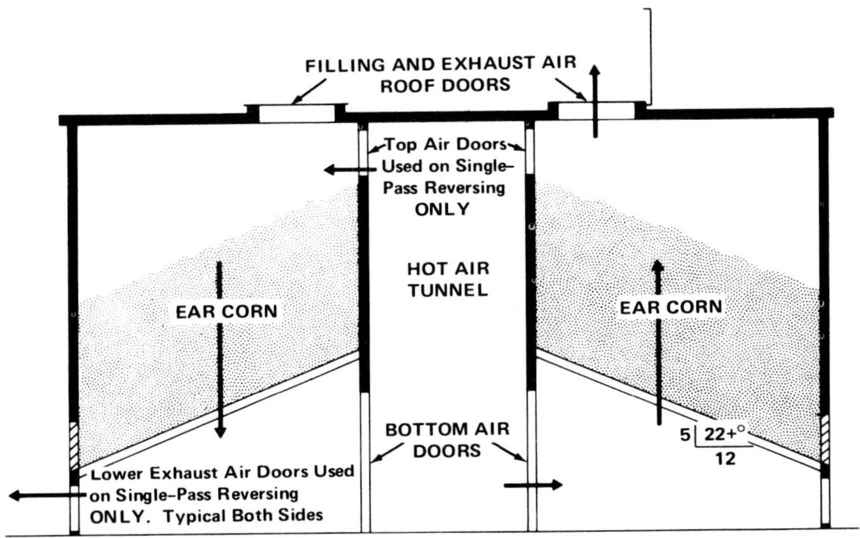

FIGURE 11.—Single-pass and single-pass reversing drying system. (Courtesy of Corn States Hybrid Serv., Des Moines, Iowa.)

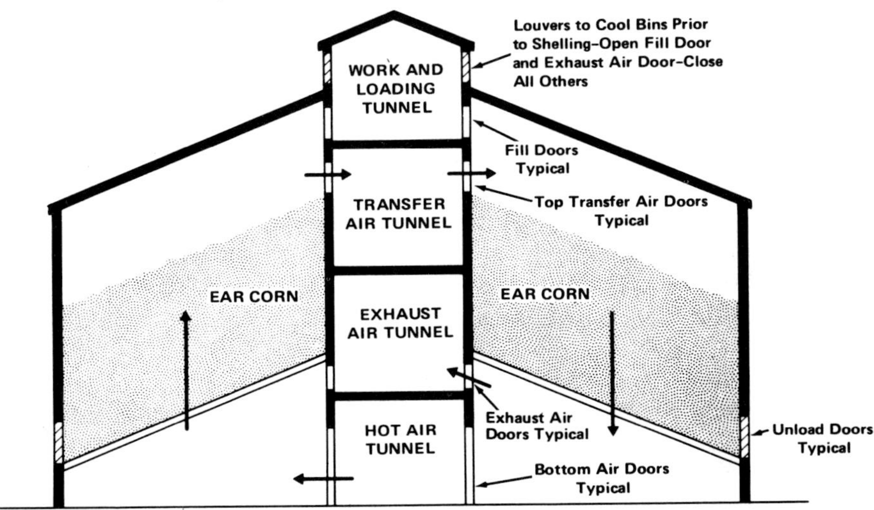

FIGURE 12.—Suction two-pass drying system. (Courtesy of Corn States Hybrid Serv., Des Moines, Iowa.)

one-pass or two-pass design and is worthy of careful consideration when a new installation is planned.

The tunnel dryer is another form of the heated air batch dryer. It is used for certain vegetable seeds, such as cucumber, muskmelon, and tomato, and for drying seeds in bags. Because vine crop and tomato seeds have to be washed to remove adhering fruit pulp, they require rapid drying immediately after washing. Rapid drying is accomplished by spreading the seeds in thin layers in screen-bottom trays, which are either placed in a tunnel dryer or dried in the sun, as explained previously. Tunnel dryers consist of open trenches through which heated air is circulated up through the seeds in the trays. Seeds in bags can be dried in a tunnel dryer by building the tunnel with a cover having special holes over which the bags of seeds are placed. This is the most inefficient system of drying seeds. A tunnel dryer can be constructed with a capacity to meet a specific need.

Continuous Flow Dryers.—The continuous flow dryer may be horizontal or vertical (fig. 13). It is called continuous flow because the seeds move slowly from the point of entry to the point of exit and are dried as they move through the unit. A continuous flow dryer is not suited for drying small quantities. For best efficiency, the system must be full of seeds at all times. Consequently, the continuous flow dryer is not at all suited to an operation requiring frequent changes from one kind or cultivar of seed to another. Its greatest value is for grain drying where thousands of bushels of a single cultivar can be dried without interruption.

Modification of Basic Dryer System.—The dryeration (fig. 14) represents a new grain drying process developed at Purdue University. It is a combination of high-speed and high-temperature drying and slow cooling (McKenzie et al., *1968*). With the dryeration process, no cooling is done in the dryer. Grain is discharged at 50° to 60° C from the dryer, still carrying 1- to 3-percent excess moisture into a separate cooling bin. The hot grain is held without mechanical cooling for a tempering period. It is then cooled slowly to remove the excess moisture before being transferred, dry and cool, to storage or load out. The dryeration process, because of the high drying temperature and the tempering period, may not be as satisfactory for seed drying as it is for grain drying.

Dehumidified Air Systems.—In dehumidified air systems, a dehumidifier dries the air before it is heated and passed through the seeds. The air dryer converts the latent heat in the moisture to sensible heat in the drying airstream, which lowers the relative humidity of the air by decreasing the total moisture in the air. The rise in temperature associated with dehydration reduces the amount of heat needed from the main heat source, which reduces fuel costs while drying efficiency is increased. Dehumidified air can be used with either a batch-type or a continuous flow dryer. All that is required is to install a dehumidifier in the air delivery system.

Dehumidification, although generally used with a closed system, can be incorporated into any drying system. The air passes through the

FIGURE 13.—Vertical continuous flow dryer. (Courtesy of Ferrell-Ross, Oklahoma City, Okla.)

dehumidification medium, usually silica gel, and then is circulated through the rest of the system back to the dehumidifier (fig. 15). Since the drying medium has to be recharged or dried out, either a two-tower or a rotary system is used. With the two-tower system (fig. 16), one tower dehydrates the air while the drying medium in the other tower is being dried. The switch from one tower to the other is automatic. With a rotary dehumidifier (figs. 17, 18) the drying agent is held in a circular unit, which continuously rotates so that one section is drying the air while the desiccant in another section is being dried.

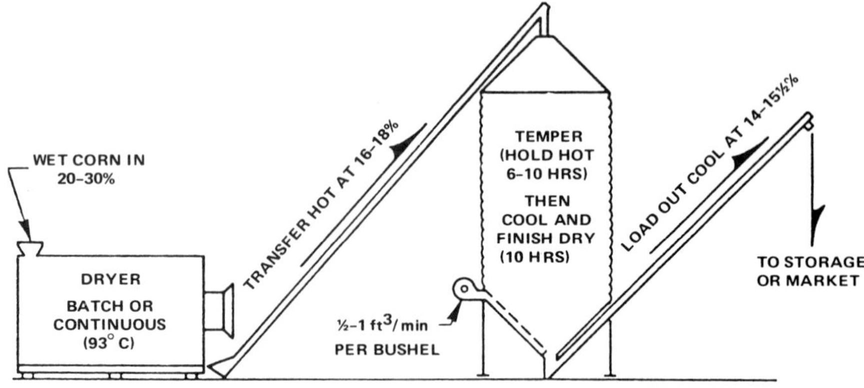

FIGURE 14.—Schematic flow diagram of dryeration process. (Courtesy of Coop. Ext. Serv., Purdue Univ.)

In general, seed drying plants are custom engineered for specific purposes (figs. 19, 20). For example, a dryer plant designed to dry ear corn cannot be used to dry Kentucky bluegrass strippings. Dryers, as well as drying plants, are sometimes designed for special purposes. A special-purpose dryer (fig. 21) was developed by the National Seed Storage Laboratory for conditioning small quantities of seed for storage research projects. The trays used in the dryer were designed for loose seed; however, by means of a system of cross wires to allow air movement between packets, the trays can be used for either bulk or packet drying, especially if the wire grid is removable.

A special walk-in seed-drying room was constructed at the New York Agricultural Experiment Station to handle relatively large samples of plant materials in bags (fig. 22) (Dolan et al., *1973*).

Miscellaneous Dryers.—Seeds can also be dried under a partial vacuum, by the freeze-drying process, or by infrared radiation (Schroeder and Rosberg, *1959*), all of which are used extensively in the food industry. Each of these drying methods has certain advantages and disadvantages. In general, they are more costly than conventional drying methods, and in some cases they are suitable only for drying small quantities of seed. In vacuum and freeze-drying, seed moisture content can be reduced to very low levels without using excessively high temperatures. Seed dried to moisture levels below air dryness must be stored in either moisture-barrier containers or dehumidified rooms; otherwise the advantage of such drying will soon be lost through rehydration.

Much of this information was reported by Anderson and Alcock *(1954)*, Hall *(1957)*, Wheeler and Hill *(1957)*, Brandenburg et al. *(1961)*, Kreyger *(1963a, 1963b, 1972)*, Barre *(1963)*, McKenzie *(1966)*, McKenzie et al. *(1968)*, Harrington and Douglas *(1970)*, Peterson *(1971)*, and Philpot *(1976)*.

Factors To Consider in Selecting a Dryer System

Drying Capacity.—The drying capacity should be sufficiently large to permit continuous, efficient harvest of the seed crop, regardless of

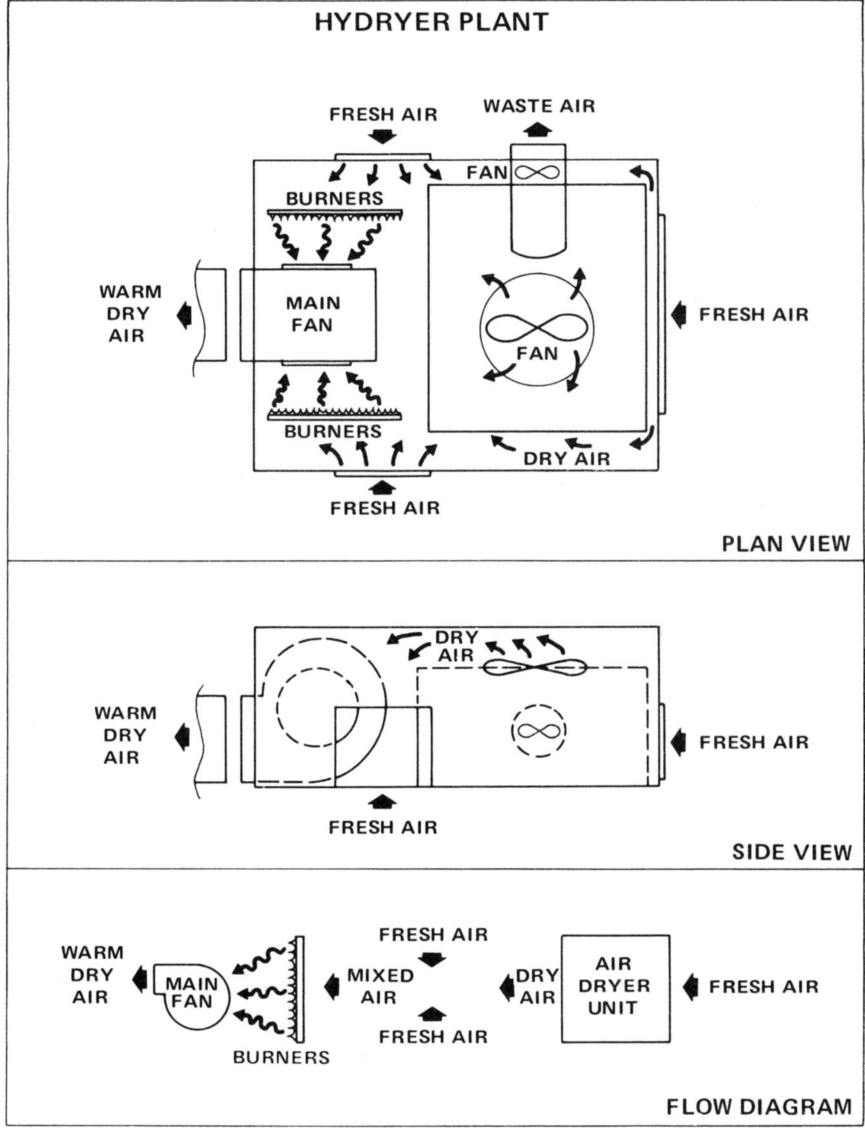

FIGURE 15.—Flow diagrams of air movement through a combination dehumidified-air heated-air dryer system. (Courtesy of Corn States Hybrid Serv., Des Moines, Iowa.)

whether only one or several cultivars are to be dried. Daily capacity of the drying system should efficiently handle one day's harvest.

Investment Necessary To Get Capacity.—If the investment in drying capacity is excessively high, costs of drying are bound to be high. Since ownership costs are often about half the cost of drying, the initial cost as well as the maintenance costs should be carefully considered when selecting a dryer.

Fuel and Power Costs.—The cost of fuel to heat the drying air and

PN–5399

FIGURE 16.—Two-tower dehumidifier for use in storehouses. (Courtesy of Seedburo Equipment Co., Chicago, Ill.)

power to operate the fan may not vary much among dryers; however, such costs must be included when total cost of drying is calculated. Since high-speed systems tend to increase costs more than capacity, they cost more to operate.

Airflow.—Efficiency of a dryer is directly related to airflow, which is determined by the output of the fan. Generally the airflow per horsepower does not vary greatly among fans of comparable size. However, some industry rating practices are confusing. For example, fan motors sometimes deliver more horsepower than their nameplate rating. Nameplates on fan motors sometimes carry a dual rating, showing a range of horsepower. Such practices tend to make one fan seem more efficient than another with the same horsepower rating when the difference is actually the result of one motor delivering a higher horsepower.

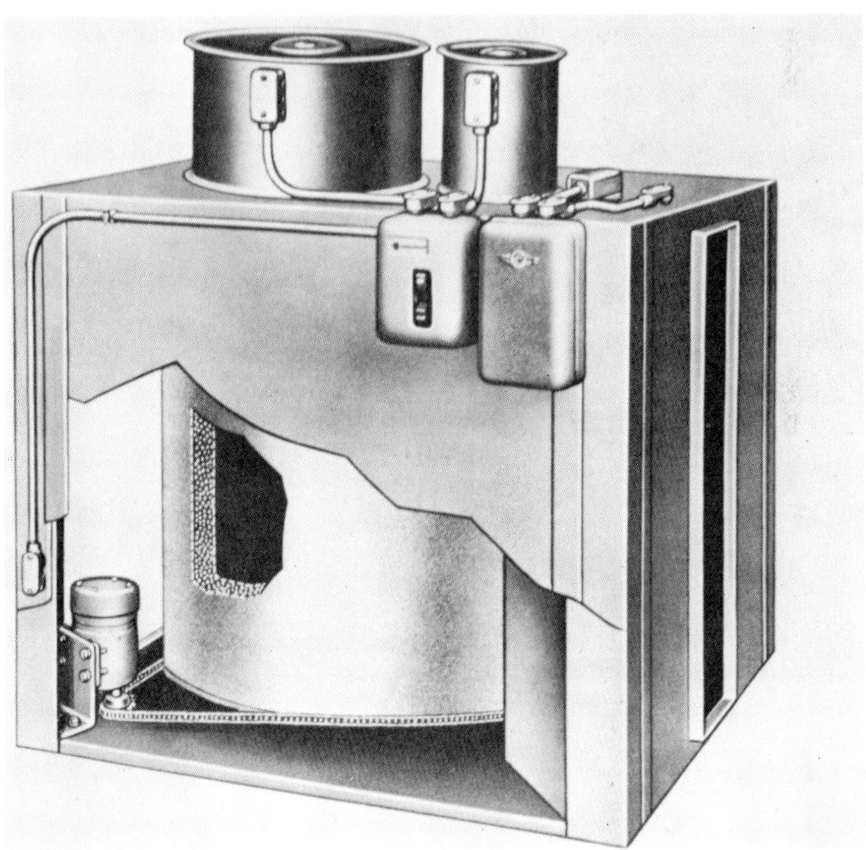

FIGURE 17.—Cutaway view of a rotary dehumidifier. (Courtesy of Corn States Hybrid Serv., Des Moines, Iowa.)

PN–5400

FIGURE 18.—High capacity rotary dehumidifier with 50,000 cubic-foot-per-minute airflow. Equipment includes a 150,000 cubic-foot-per-minute capacity main blower and air-heater unit. (Courtesy of Corn States Hybrid Serv., Des Moines, Iowa.)

PN–5401

FIGURE 19.—This corn drying plant utilizes the most recent advancements in drying and handling equipment. (Courtesy of Equipment Specialists, Inc., Taylorville, Ill.)

PN–5402

FIGURE 20.—The dried ear corn from figure 19 passes from the bins along the permanent discharge conveyor directly into the sheller. (Courtesy of Equipment Specialists, Inc., Taylorville, Ill.)

Heat Input and Drying Temperature.—Assuming the airflow is sufficient for the drying method used, failure of a system is usually the result of heat management. Systems fail because of too much or too little heat. Too much heat reduces seed quality, and too little heat slows drying. There are desirable and workable air-heat ratios for each dryer system. These ratios vary with the dryer and the kind of seed. For best quality, seed temperature should not exceed 43° C for grain and 32° for vegetable seeds. Higher temperatures will increase drying rate but will also reduce seed quality. Air temperature can exceed seed temperature if the seed is not exposed long enough to reach air temperature or pass critical limits. Table 19 gives effects of heating time, temperature, and seed moisture content on loss of viability for several kinds of seed.

Laboratory studies by Bass *(1953)* and Grabe *(1957)* showed that the higher the moisture content of Kentucky bluegrass and smooth bromegrass seeds, the lower the temperature and the shorter the time at that temperature required to bring about viability or germination loss (fig. 23) (table 20). Grabe *(1957)* also showed the effects of various temperatures and drying periods on the germination of smooth bromegrass seeds (table 21). He noted that none of the temperature-time treatments used adversely affected the germination of smooth bromegrass seeds, even after 29 months of storage.

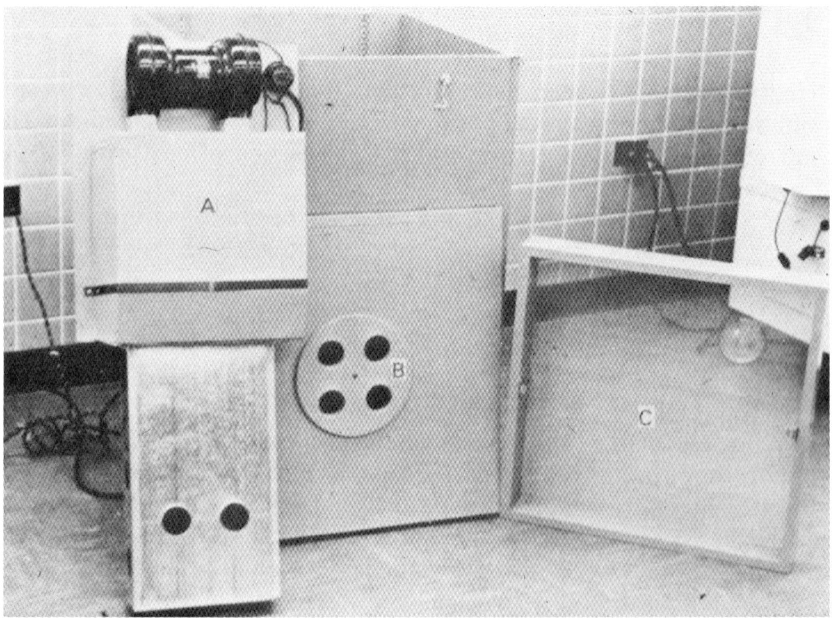

PN–5403, PN–5404

FIGURE 21.—Seed dryer developed by National Seed Storage Laboratory, Fort Collins, Colo. *Above*, assembled; *below*, disassembled, showing twin-blower air diffuser *(A)*, top ventilation holes *(B)*, and screen-bottom seed tray *(C)*.

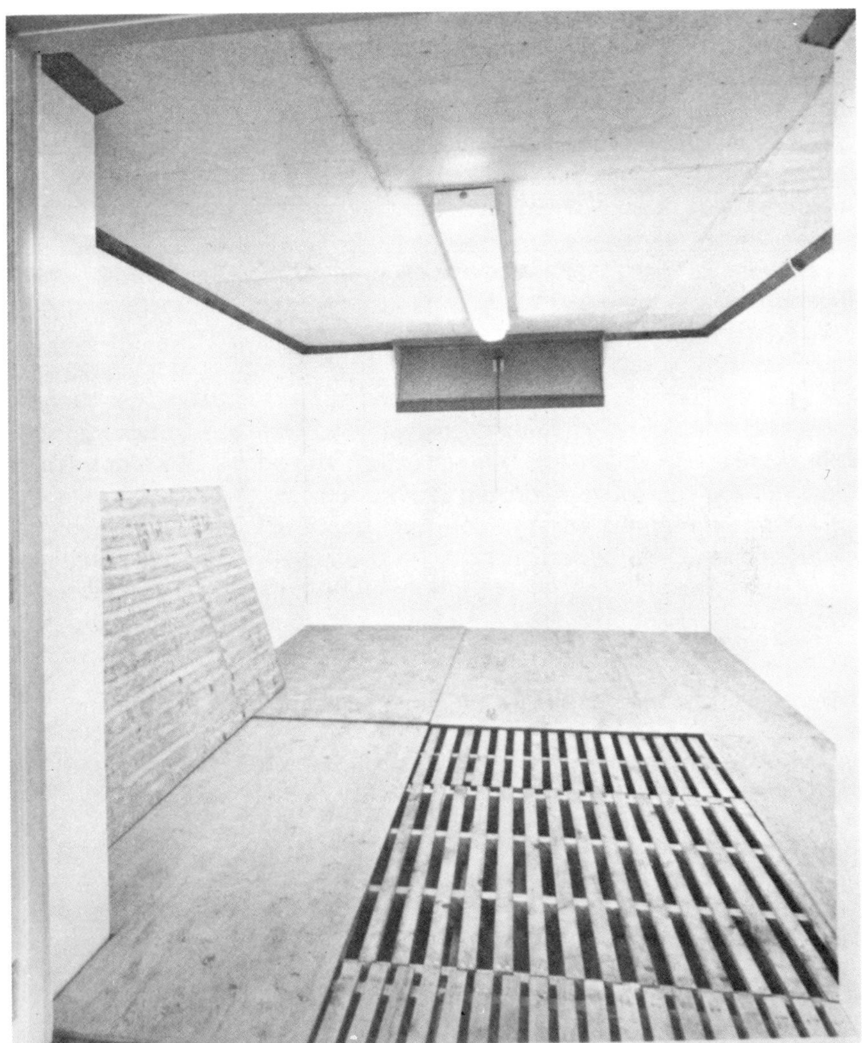

PN–5405

FIGURE 22.—Interior of a seed-drying room, showing slotted floor, plywood covers, exit ducts, and inlet hot air duct near ceiling. (Courtesy of N.Y. Agr. Expt. Sta., Geneva.)

Williams *(1938)* reported that it is advisable to remove excess moisture from perennial ryegrass, timothy, and orchardgrass (cocksfoot) seeds as quickly as possible after harvest. He referred to excess moisture as being that above the norms he set. They were about 14 percent for ryegrass, 13.6 percent for timothy, and 13 percent for orchardgrass. Because of their greater rate of absorption and slower rate of drying, seeds of perennial ryegrass require more careful atten-

TABLE 19.—*Effect of different heating times on viability of various seeds at different seed temperatures and moisture contents (static conditions)*[1]

Seed	Moisture content	Effect of heating time on seed viability[2] at indicated seed temperature (°C)				
		30	40	50	60	70
	Percent	*Hours*	*Hours*	*Hours*	*Hours*	*Hours*
Broadbean	12	4-	4-	2-	1½-	¼*
	16	4-	4-	2-	¼*	1/6*
	24	3-	3-	1*	1/6*	1/6*
Lupine	10	4-	4-	2-	1½*	1-
	14	4-	4-	2-	1½*	½*
	19	3-	3-	2-	½*	1/6*
Maize	12	4-	4-	2-	¼*	1/6*
	15	4-	4-	½*	¼*	1/6*
	21	3-	3-	½*	¼*	1/6*
Oats	12	4-	4-	2-	1½-	¼*
	14	4-	4-	2-	¼*	1/6*
	19	3-	3-	½*	¼*	1/6*
Pea	12	4-	4-	2-	1½*	¼*
	15	4-	4-	2-	1½*	¼*
	21	3-	3-	½*	1/6*	1/6*
Rape	6	4-	4-	2-	1½-	1-
	9	4-	4-	2-	1½-	¼*
	13	3-	3-	2*	½*	1/6*
Rye	13	4-	4-	2-	¼*	1/6*
	16	4-	4-	½*	¼*	1/6*
	21	3-	3-	½*	¼*	1/6*
Ryegrass	12	4-	4-	2-	1½-	1-
	15	4-	4-	2-	1½*	¼*
	22	3-	3-	2*	¼*	1/6*
Sugar beet	11	4-	4-	2-	½*	¼*
	15	4-	4-	2-	¼*	¼*
	20	3-	3-	2-	¼*	1/6*

[1] Data from Kreyger *(1963)*.
[2] Asterisk (*) indicates decrease and dash (-) no decrease in viability.

tion during drying than those of orchardgrass and timothy. A current of warm air at not more than 43° C alternating with a cool (15°–18°) one (20 min warm, 10 min cool) is particularly beneficial for drying ryegrass and orchardgrass seeds. Drying with a continuous warm air current adversely affected viability, even when the temperature did not exceed 35°.

Wileman and Ullstrup *(1945)* found that germination of ear corn with an initial moisture content of 35 percent or more dropped rapidly when dried at 49° C; however, corn with 25 percent or less initial moisture suffered no appreciable germination loss when dried at 49°, whereas corn with 20 percent or less initial moisture showed no germination loss when dried at 54°.

Drying at too high a temperature or too rapidly can drastically reduce seed viability (Toole and Toole, *1946*; Griffith and Harrison, *1954*; Raun

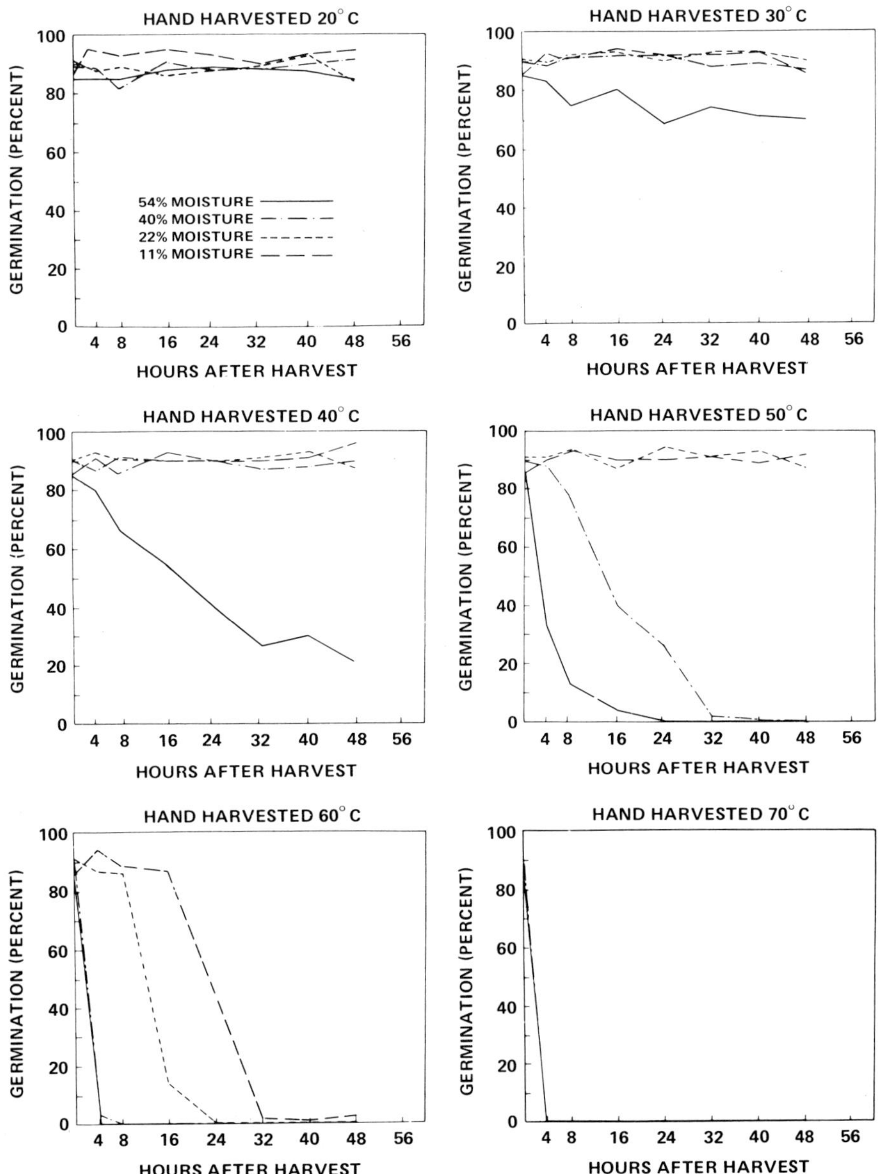

FIGURE 23.—Germination of Kentucky bluegrass seeds with four moisture contents after 4–48 hours at various temperatures without drying.

TABLE 20.—*Viability of bromegrass seed as affected by interactions of temperature, moisture content, and duration of treatment in 1953 and 1954 laboratory studies*

Year, temperature, and moisture content (percent)	Viability after indicated hours of heat treatment								
	0	4	8	12	20	28	36	44	52
	Percent	*Percent*	*Percent*	*Percent*	*Percent*	*Percent*	*Percent*	*Percent*	*Percent*
1953									
40° C:									
56.6	88	89	84	87	91	89	88	90	87
51.4	93	91	90	92	94	90	93	91	88
39.3	83	94	90	92	89	89	85	88	86
32.4	89	99	97	94	93	96	94	91	92
20.0	95	93	94	92	97	95	95	93	95
13.2	93	96	94	90	90	96	96	86	91
50° C:									
56.6	88	40	5	0	0	0	0	0	0
51.4	93	79	30	12	3	0	0	0	0
39.3	83	90	72	66	34	6	3	1	1
32.4	89	93	89	67	75	49	32	11	5
20.0	95	91	92	90	91	92	89	73	89
13.2	93	94	95	93	94	99	95	87	93
60° C:									
56.6	88	1	1	0	0	0	0	0	0
51.4	93	4	0	0	0	0	0	0	0
39.3	83	41	2	2	0	0	0	0	0
32.4	89	69	27	29	0	0	0	0	0

20.0	95	84	85	84	85	1	0	0	0
13.2	93	91	95	93	93	94	97	94	88
1954									
40° C:									
38.1	91	---	---	---	---	---	---	---	85
28.4	89	---	---	---	---	---	---	---	90
15.0	91	---	---	---	---	---	---	---	93
50° C:									
38.1	91	63	84	85	38	19	21	12	8
28.4	89	90	89	82	77	79	12	0	0
15.0	91	96	91	90	90	93	92	90	90
60° C:									
38.1	91	20	0	0	0	0	0	0	0
28.4	89	0	0	0	0	0	0	0	0
15.0	91	91	88	73	72	70	63	34	0

TABLE 21.—*Summary of results of artificial drying experiments on germination of 13 lots of smooth bromegrass seed in 1953 and 1954*[1]

Year and lot size (lb)	Drying temperature	Temperature of seed[2]	Drying time	Moisture content		Germination	
				Before	After	Before	After
	°*C*	°*C*	*Hours*	*Percent*	*Percent*	*Percent*	*Percent*
1953							
Unknown	41	40	4¼	20.0	13.0	81	83
Do	49	30	4	20.2	14.9	91	90
Do	52	41	3¼	17.0	11.4	94	93
Do	55	55	3¾	21.0	12.0	86	90
Do	60	65	1¼	21.3	12.3	92	91
Do	[3]43, 52	48	2, 2	20.8	13.3	84	91
Do	[3]49, 60	50	2, 2½	28.8	17.0	90	91
Do	[3]49, 66, 71	56	2, 1, 2¼	27.2	16.0	90	89
1954							
917	60	64	1	14.5	6.7	89	90
900	66	62	2	23.9	5.9	91	90
2,100	77	76	2	19.4	7.7	93	96
2,400	88	84	2	12.1	5.0	93	92
5,600	99	85	4	14.3	- - -	95	96

[1] Data from Grabe (*1957*).
[2] Maximum temperatures in bottom 6 inches; temperatures in remainder of load were lower than these data.
[3] Seed dried 2 hours at lower temperature, then completed at higher temperatures.

and Frey, *1957*). Excessive drying can also result in loss of viability and the development of seedling abnormalities, especially under certain storage conditions (Evans, *1957b*; Nutile, *1964a, 1964b*). Heated air-drying may induce a dormant condition, which can seriously affect field stand establishment unless it is overcome before planting (Wright and Kinch, *1962*; Nutile and Woodstock, *1967*).

Handling and Labor Requirements.—The various drying methods obviously differ in the amount of handling equipment and labor needed. In-storage layer drying involves two seed handling operations—filling and unloading the bin. Batch and continuous flow drying require four operations, namely, loading and unloading the dryer and filling and unloading the bin. The volume of seed to be dried determines handling equipment cost rather than the type of dryer. A large volume of seed requires a more elaborate and expensive seed handling system.

Weather.—Weather is a major factor in seed drying operations. It determines the amount of time available for harvesting and drying a seed crop. Extended periods of cool, damp weather can markedly increase harvesting and drying costs, whereas hot, dry winds can reduce such costs. During periods of wet weather, extra heat and air movement are required for efficient drying. During periods of hot, dry weather, the amount of supplemental heat needed for rapid drying is much less than that required during warm, humid weather.

Drying Rate.—The rate at which moisture leaves a seed depends on how much the seed lacks moisture equilibrium with the air surrounding it, the air temperature, and the composition, size, and shape of the seed. When the initial seed moisture content is high, the rate is faster if the temperature is high or if relative humidity is low. A change from very slow to rapid air movement will increase the rate of drying. The rate of drying drops off as the moisture content of the seed decreases. This means that as the moisture content of seed decreases, longer exposure to heat is required for each percentage of moisture removed.

In bulk drying the rate of moisture loss is not uniform throughout the depth of seeds in a dryer. As the seeds dry, the interface between dry and wet seeds gradually moves through the seed layer. Since the seeds nearest the heated air source dry fastest and those farthest away dry slowest, there is always a range in seed moisture levels through the layer of seeds in a dryer.

The drying rate for individual kinds of seeds depends on their individual properties as to release of moisture. However, various kinds can be grouped with regard to their tendency to release moisture (Kreyger, *1963a*). There are fast drying seeds such as grasses, rape, and sugar beet; normal drying seeds such as barley, oats, rice, rye, and wheat; and slow drying seeds such as bean, corn, lupine, and pea (table 22).

TABLE 22.—*Comparison of approximate rates of drying and releasing moisture at 5 moisture contents for various seeds*[1]

Seed and drying group	Rates at indicated moisture content (percent) - wet basis				
	22	20	18	16	14
	DRYING PER °C PER HOUR				
	Percent	*Percent*	*Percent*	*Percent*	*Percent*
Broadbean	0.14	0.08	0.05	0.03	0.016
Garden pea (green)	.19	.11	.07	.045	.03
Lupine	.33	.19	.11	.07	.04
Maize	.27	.18	.10	.06	.03
Oats	.9	.5	.3	.2	.12
Rape	- - -	3	2.3	1.6	1.0
Rye	.45	.35	.23	.14	.07
Ryegrass	1.5	.8	.5	.32	.2
Sugar beet	3.3	2.4	1.6	1.0	.6
Wheat	.42	.3	.2	.13	.07
	RELEASING MOISTURE (TENDENCY)[2]				
Slow drying	*Number*	*Number*	*Number*	*Number*	*Number*
Broadbean	70	40	25	15	8
Garden pea (green)	95	55	35	23	15
Lupine	165	95	55	35	20
Maize	135	90	50	30	15
Normal drying					
Oats	450	250	150	100	60
Rye	225	175	115	70	35
Wheat	210	150	100	65	35
Quick drying					
Rape	- - -	1,500	1,150	800	500
Ryegrass	750	400	250	160	100
Sugar beet	1,650	1,200	800	500	300

[1] Data from Kreyger *(1963a)*.

[2] Wheat at 18-percent moisture content is taken as 100; all values are relative to this norm.

Williams *(1938)* found that by using a 20-minute warm air (43° C) and 10-minute cool air (15°–18°) alternation, moisture content of ryegrass, timothy, and orchardgrass seeds can be reduced about 4.5 percent in 1½–2 hours. Drying time for 25 percent or lower moisture ear corn can be reduced about 20 percent by using 49° rather than 43° (Wileman and Ullstrup, *1945*).

The average time required to dry six lots of 21- to 25-percent moisture ear corn to 12- to 13-percent moisture was reduced from 81½ to 64½ hours by raising the drying air from 43° to 49° C and thereby decreasing its relative humidity. Drying rates for specific kinds of seed are largely unavailable because such factors as the initial moisture content, the temperature, relative humidity, and velocity of the air, and the depth of the seed layer must all be considered simultaneously.

A preliminary estimate of drying rate can be obtained by recording the moisture loss from a known weight of seed held under known drying conditions for a specified length of time. Knowledge of whether the kinds of seeds are slow, normal, or fast drying is very helpful in determining the type of dryer system to install. For example, a continuous flow dryer is generally suitable for quick and normal drying seeds but not for slow drying seeds, nor is it suitable for nonfree-flowing seeds. Batch dryers are better for slow drying and nonfree-flowing seeds. A low temperature is used at the start and gradually increased as drying progresses. This practice is fairly common in the field corn seed-producing areas where double-pass dryers are used.

Drying Seeds With Desiccants

Barrow *(1915)* reported that 5-percent calcium chloride by weight minimized heating and deterioration of cottonseeds during storage. Kondo and Isshiki *(1936)* reported that either calcium oxide or calcium chloride could be used to desiccate rice seeds. They found that calcium oxide absorbs only about one-third as much moisture as calcium chloride but absorbs it faster. One kg of calcium chloride or 3 kg of calcium oxide will absorb 1-percent moisture from 5.1 bushels of rice. According to Kondo and Terasaka *(1936)*, both calcium chloride and calcium oxide desiccate rice for safe, sealed storage for food but not for seeds. Dexter and Creighton *(1948)* found that wood blocks impregnated with calcium chloride, magnesium chloride, magnesium sulfate, or sodium chloride effectively removed moisture from stored grain; however, they did not determine what effect, if any, storage with the impregnated wooden blocks had on germination.

Although seeds can be dried safely with desiccants, drying large quantities of high moisture seeds with desiccants is neither practical nor economically feasible because of the large amount of desiccant needed to remove 1 percent of water from each pound of seed. Use of desiccants with small quantities of seed in sealed containers is discussed under "Packaging and Packaging Materials."

SEED STORAGE STRUCTURES

A considerable amount of seed becomes useless each year from improper storage. Seeds must be stored dry and kept dry. As previously indicated, longevity of seeds is controlled primarily by seed moisture content and storage temperature. For maximum storage life, these two conditions must be carefully controlled. However, much of the seed produced each year needs to be stored only from harvest until the next planting season. Such seed may or may not require storage under conditions other than normal air temperature and relative humidity depending on the kind of seed and the local climate. A seed storage facility must have certain basic features, regardless of any adaptations for special purposes (Barre, *1954*; Wheeler and Hill, *1957*).

Basic Features of Storage Structures

Protection From Water

Rain, snow, ground moisture, or any source of water should never be allowed to come in contact with seeds, as it increases seed moisture content. High seed moisture content increases respiration, heating, and mold growth and sometimes promotes sprouting in storage, all of which decrease seed quality. The roof and sidewalls of seed storage structures must be free of holes and cracks that permit the entry of rain or snow. Cracks and knotholes in wood walls and roofs should be filled. All bolts and screws in metal buildings should have rubber washers, especially those in the roof. Since soil moisture can be absorbed readily on contact by seeds, all seed storage structures must have a waterproof floor. A wood floor should be elevated and a concrete floor should have a moisture barrier under it. A metal floor is naturally impermeable to water but will soon rust out if kept moist continually.

Protection From Contamination

Because individual seeds of different cultivars of various crops are indistinguishable from each other and because the kinds of seeds are often difficult to separate, each seed lot must be kept free of seeds from other lots. Storage facilities should be constructed to provide maximum protection from chance contamination. For bulk storage, a separate bin must be provided for each cultivar. For bag storage, seeds of each cultivar must be stacked separately. Seeds may also be stored in toteboxes. Toteboxes and stacks of bags on pallets can be moved readily by a forklift. All bags, toteboxes, and bins must be carefully and conspicuously labeled.

Protection From Rodents

Since large quantities of seeds may be lost because of inadequate protection from rodents, precautions must be taken to prevent their entry into storage buildings. Metal and concrete buildings normally provide good protection from rodents. With care, wood buildings can also be constructed to keep them out. Metal bins with tight covers provide rodent protection, and cloth bags can be treated to repel rodents. For further information, see the section on "Effects of Pests and Chemicals on Seed Deterioration in Storage."

Protection From Insects

Although insects do not constitute a major problem with some kinds of seeds, a good storage facility should be so constructed that any part, or all of it, can be fumigated to control insects at any time. Insect problems can be kept to a minimum by thoroughly cleaning and fumigating each time a bin is emptied. Areas where bags and toteboxes are stored should be kept free of loose seeds and trash at all times. Cleanliness eliminates insect breeding places and facilitates insect control. For further information, see "Effects of Pests and Chemicals on Seed Deterioration in Storage."

Protection From Fungi

Under certain circumstances, storage fungi can cause considerable damage to stored seeds. Since fungi grow best under warm, humid conditions, storage structures should provide cool, dry conditions. Damage from storage fungi can be kept to a minimum by carefully drying seeds to a safe moisture content, usually less than 12 percent, and holding them under dry conditions. Sometimes ventilation of storage structures, especially steel buildings, is necessary to prevent the accumulation of translocated moisture. Temperature differentials can cause water vapor to move from warmer to cooler areas of a bin, usually to the upper surface. Such moisture movement can provide conditions favorable for fungus growth unless adequate precautions are taken to prevent its accumulation. Seeds can be treated with various chemicals to control fungi. Fungicides are applied routinely to many kinds of seeds as part of their normal processing. However, fungicides to control soil fungi may not control storage fungi. For further information, refer to "Effects of Pests and Chemicals on Seed Deterioration in Storage."

Protection From Fire

Danger from fire loss is greatest with wood buildings; however, wood can be chemically treated to retard burning. The hazard of a fire in wood buildings can be reduced by cleanliness, both inside and around the building. All seed storage buildings, especially processing areas, should be equipped with special dustproof and sparkproof electrical outlets and switches, which greatly reduce the chances of an electrical fire. All wiring should be rodentproof. Although metal and concrete buildings are fireproof, the same electrical wiring precautions should be practiced in them as in wood buildings because dust explosions and fires can be caused by electric sparks. Structures with these general characteristics should provide safe storage for most kinds of seeds from harvest to the next planting season, except in areas with extremely high temperatures and relative humidity. In such areas, additional protective measures, namely temperature and humidity control, are required to maintain seed quality during storage.

Types of Storage Structures

Farm Storage

Farm storage is usually for a few days or weeks during the harvest season. Seeds are seldom held on the farm longer than from harvest to the next planting season. They may be stored in bins, bags, or in some cases on the ground. Ground storage is generally temporary and used only in an emergency. Bulk lots may be stored in wood or metal bins, granaries, and barns. Seeds, unless very dry, keep better in wood than in metal buildings because heat buildup is greater and moisture loss is slower in metal than in wood buildings. Heat accumulation in metal buildings can seriously reduce the viability of seed stored with high moisture. Heat buildup can be reduced by painting the outside of the building white, by ventilation, or both. Seeds in bags may be stored in any farm building where space is available. This practice frequently causes large losses through insect and rodent damage. When bagged with a high moisture content, seed will heat and soon become worthless. (Barre, *1954*; Wheeler and Hill, *1957*; Harrington and Douglas, *1970*)

Country Elevator Storage

Many country elevators accumulate seeds during and immediately after harvest. Storage, which is usually for a limited time, may be either in bulk, in wood or metal bins, in concrete silos, or in bags or toteboxes in wood, brick, concrete, or metal buildings. Country elevators frequently have a fanning mill to remove trash before storing seeds. Fanning also helps reduce seed moisture content. Country elevators frequently have facilities for insect and rodent control and may have drying equipment (Barre, *1954*).

Seed Processor Storage

Seed processors as a group store a major part of the seeds held for future use in the United States. Their storage facilities vary widely in size and construction, ranging from small wood or metal buildings to large multistoried wood, brick, or concrete warehouses. Some processors use well-insulated rooms, in which the temperature and relative humidity are controlled for maximum viability protection, especially for high-cost items. Seed processors store both processed and unprocessed seeds. Uncleaned seeds may be put in bags, pallet boxes, or bulk bins. Each container has certain advantages and disadvantages for specific kinds of seeds. Uncleaned seeds are usually stored in an area well separated from where the cleaned seeds are stored. Completely processed seeds are generally stored in the container in which they will be shipped. Some processed seeds are stored by the processor in retail packages, but other seeds are repackaged by the retailer. Storage facilities operated by seed processors usually have the basic requirements previously discussed and may also provide for special needs of individual kinds of seeds and for protection from unfavorable climatic conditions.

Retail Storage

Seed retailers must hold their stock for varying lengths of time, usually under normal atmospheric conditions for their individual locations. Retail storage is usually far from ideal; however, except in very hot, humid areas, the shelf life of seeds in cloth, paper, or thin plastic containers is usually sufficiently long to permit marketing with little loss of viability. In hot, humid areas, shelf life of seeds is limited unless they are dried to a safe level of 5- to 8-percent moisture and put in moisture-barrier packages. At the retail level little or no attempt is made to provide protective storage.

Research Storage

Scientists must store varying quantities of specific seed stocks in order to have viable seeds available for their research. Although refrigerated, dehumidified storage is best suited for long-term holding of seeds, such conditions are not uniformly available to research scientists. They frequently must use less than ideal conditions, although ordinarily they use the best facilities available.

In some geographic areas, numerous kinds of seeds can be stored without loss of viability for several years under normal atmospheric conditions. In tropical and subtropical areas, controlled atmospheric storage conditions are essential to preserve research seed stocks for periods longer than a few days. The only alternative method is sealed storage. For safe, sealed storage for up to 3 to 5 years at ambient temperatures, seeds must be dried to 5- to 8-percent moisture content before they are sealed in moistureproof containers. For longer storage, the moisture content must be reduced to 2.5 to 5 percent before packaging.

Germ Plasm Storage

Germ plasm preservation centers, such as the U.S. National Seed Storage Laboratory (fig. 24) (James, *1972*), the U.S. Regional Plant Introduction Stations, the Japanese National Seed Storage Laboratory (Ito, *1972*), and similar facilities throughout the world, must have the best possible storage conditions to minimize regrowing of seed stocks. Many kinds of seeds held under the conditions used in the U.S. National Seed Storage Laboratory do not have to be regrown more frequently than every 20 or 30 years.

This laboratory is a three-level building. The machine room (fig. 25), control room, biochemistry laboratory, growth chamber room, custo-

PN–5406

FIGURE 24.—U.S. National Seed Storage Laboratory, Fort Collins, Colo.

PN–5407

FIGURE 25.—Machine room of the U.S. National Seed Storage Laboratory, showing some of the equipment required for a refrigerated storage facility.

dian's supply room, workshop, garage, and a supply storage room are on the first level. The second level houses the administrative offices. The seed storage rooms (fig. 26) and germination laboratory (fig. 27) occupy the third level.

The seed storage rooms, accessible from a common corridor, have a total capacity of approximately 180,000 pint cans for samples. Seed sample capacity can be increased manyfold by using containers tailored to fit the sample. Storage conditions are maintained at about 4° C, with an average relative humidity of approximately 35 percent. This combination was selected as suitable for storing most kinds of seeds for a relatively long time. Three of the rooms are equipped to maintain a temperature as low as −12° if desired. For research purposes a variety of temperatures and relative humidities are available in smaller rooms not used for routine storage of germ plasm.

Constructing Controlled Atmosphere Seed Storage Facilities

Safe storage of seeds requires careful control of both the temperature and the relative humidity of the storage area. They cannot be controlled except in specially constructed rooms or buildings. Because of the need

PN–5408

FIGURE 26.—Sample storage room of the U.S. National Seed Storage Laboratory, where a scientist is checking samples stored in tin cans.

for effective barriers from outside sources of heat and moisture, the walls, ceiling, and floor of a seed storage room must have satisfactory heat insulation and a moisture vapor seal. Figures 28–30 illustrate construction techniques that will provide the necessary heat and moisture barriers.

Floor insulation is frequently installed in a bed of hot asphalt, which provides a good vapor seal. The amount of insulation used depends on the temperature to be maintained and the type of material used, such as fiber glass, spray on foam, Styrofoam, and cork. Insulating materials

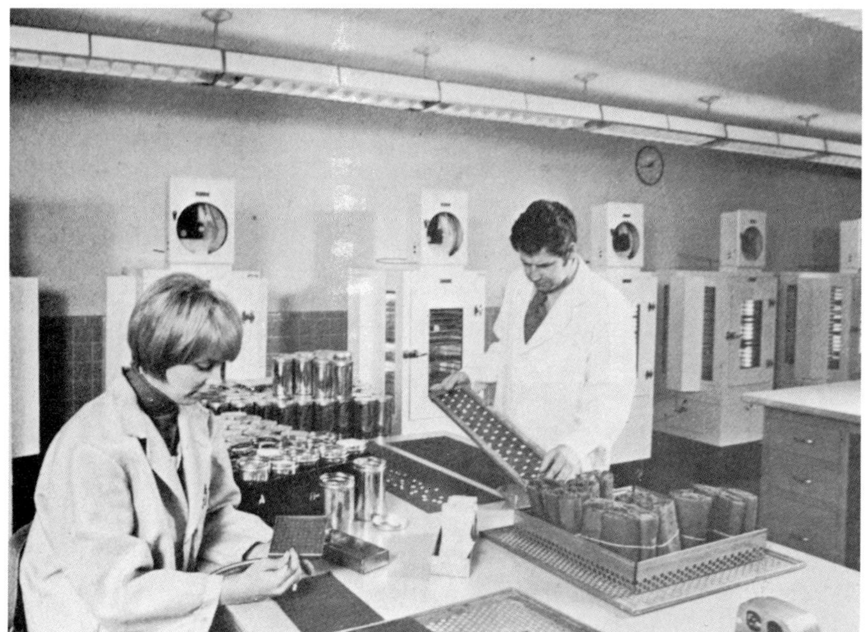

PN–5409

FIGURE 27.—Laboratory workers in the U.S. National Seed Storage Laboratory testing seeds for germination or decline in viability as storage time is increased.

must be kept dry for maximum efficiency. If the material does not have a characteristic for dryness built into it, moisture protection must be provided outside the insulation.

Board-type insulation should be applied in two or more layers, with the joints lapped or staggered to minimize heat and moisture penetration through the joints. To cope with the problem of building movement with changing temperature, an accordianfold is used in the corner flashing vapor seal material. Ceiling insulation can be of many kinds. Ceiling and wall finishes usually consist of one-half inch or more of cement plaster applied as two coats. Where the wall is subject to shock, the finish coats are reinforced with galvanized metal lath. Wood, metal, or concrete bumpers are installed on walls where trucks might accidentally hit them.

Cold storage rooms must have no windows and their doors must be well insulated and well sealed. For large openings, roller hung doors may be better than swinging doors. Roller doors not only fit tighter but can be operated electrically. A relatively new idea is the use of a high velocity stream of cool air across the face of the door, usually from top to bottom. This may not be the complete solution to the entrance of heat and moisture, but it does provide some protection. Double-door air locks

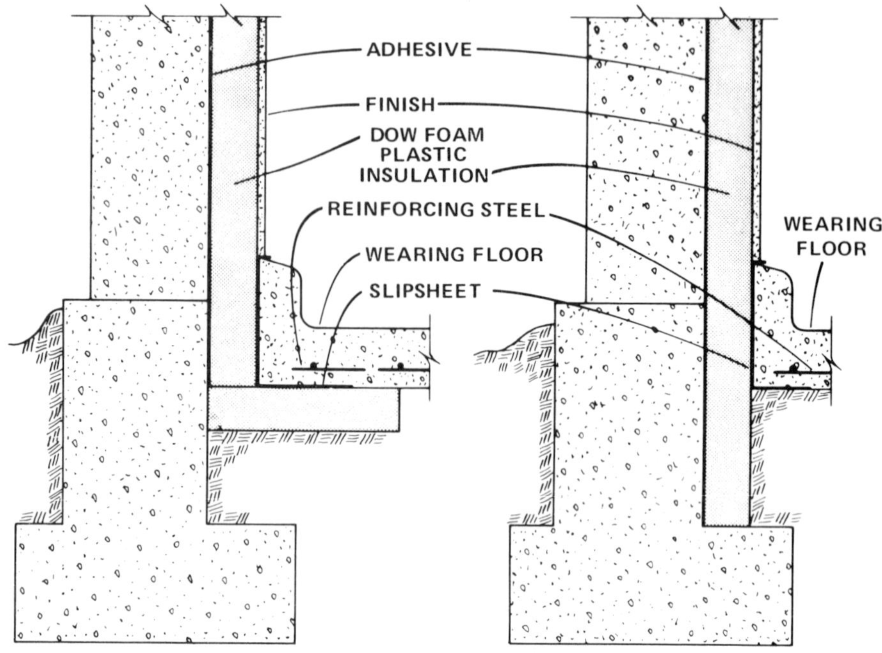

FIGURE 28.—Floor construction of a refrigerated seed or grain storehouse, emphasizing importance of tight seals and insulation. (Courtesy of Amspec, Inc.)

and small anterooms also help reduce heat and moisture entering cold storage rooms.

The biggest problems in refrigerated dehumidified storage are heat and moisture leakage through the walls, roof, floor, and around the doors; heat generated by the lights; and heat and moisture generated by people working in the room. These, of course, can best be reduced by incorporating adequate preventative measures into actual construction of the storage room or warehouse. It is usually desirable to construct several controlled temperature rooms rather than a single large warehouse. By having several individually controlled rooms, annual operating costs can be lowered significantly. During periods when only small quantities of seeds are stored, one or two rooms rather than an entire warehouse can be refrigerated.

Most refrigerated seed storage facilities use forced air circulated through a cooling coil, then throughout the room. For large areas, a duct system distributes the cold air uniformly throughout the room. The final decision as to the structural design of the building or room and the size and type of refrigeration system must be left to a competent refrigeration engineer. It is far better and more economical to install initially adequate insulation, moisture protection, and refrigeration

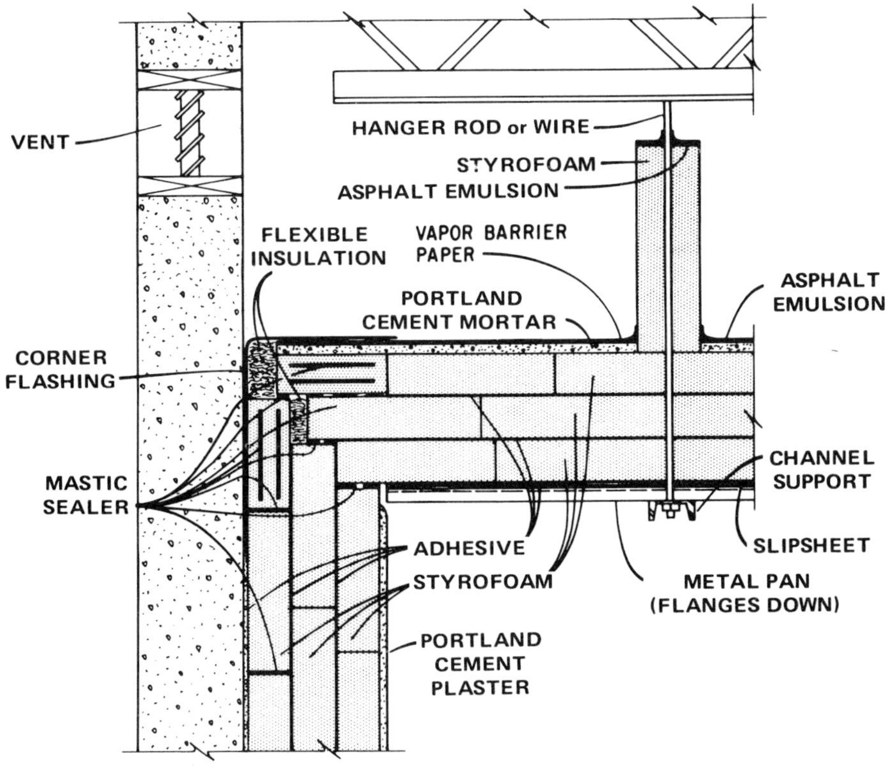

FIGURE 29.—Metal pan ceiling for blast freezer, emphasizing supporting structure, close fitting seals, and insulation. (Courtesy of Amspec, Inc.)

capacity than to remodel an unsatisfactory system. For further information, see May *(1964)*, Dryomatic Division, LogEtronics *(1965)*, Munford *(1965)*, Cooke *(1966)*, and Harrington and Douglas *(1970)*.

Controlling Temperature

Refrigeration

Since refrigeration in broad terms is any process by which heat is removed, it is the only way to obtain and maintain the low temperature required for long-term storage of seeds. Ventilation can be considered to be refrigeration; however, ventilation is suitable for only minor downward adjustments of temperature. To obtain very low temperatures, mechanical refrigeration must be used in addition to air circulation. Refrigeration is accomplished by transferring heat from the body being refrigerated to another body where the temperature is below that of the body being refrigerated. Because heat moves readily from an

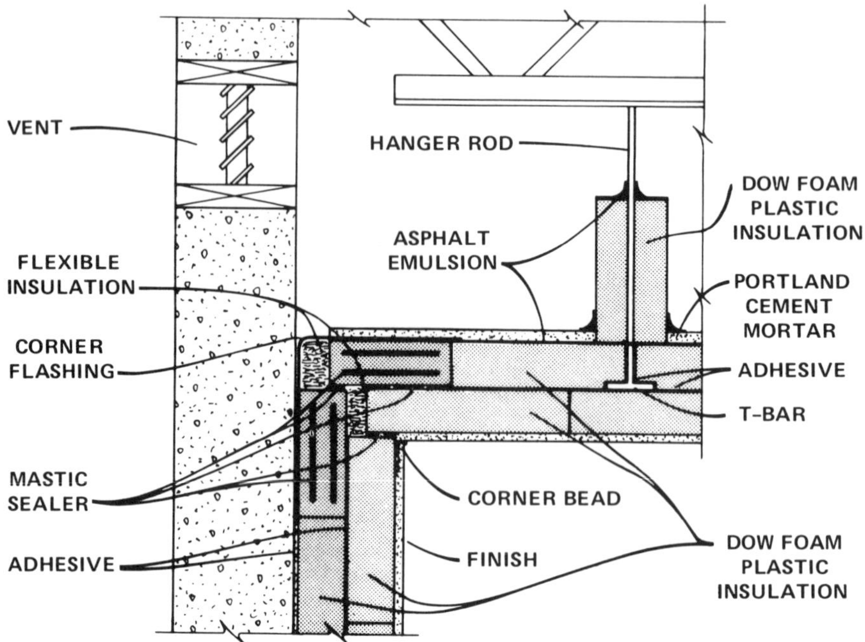

FIGURE 30.—Construction utilizing suspended T-bar ceiling and foam plastic insulation for seed storage facility. (Courtesy of Amspec, Inc.)

area of high temperature to one of lower temperature, thermal insulation is always necessary around an area to be refrigerated.

Heat Load

The heat load is the rate at which heat must be removed in order to produce and maintain the desired temperature. Heat load is affected by such factors as temperature of the space or material to be refrigerated; heat leakage through the walls, floor, ceiling, and door opening of the chamber; heat from lights, motors, and other electrical equipment; and people working inside the refrigerated chamber. Total heat from these sources determines the heat load for a particular refrigeration system.

Refrigeration Agent

The body employed to absorb heat is the refrigeration agent or refrigerant. Cooling systems are classified as either sensible or latent, according to the effect the absorbed heat has on the refrigerant. The cooling process is said to be sensible when the absorbed heat increases the temperature of the refrigerant and latent when the physical state of the refrigerant is changed. With either process the temperature of the refrigerant must always be lower than that of the area or material being refrigerated. Continuous refrigeration can be achieved by either a sensible or a latent cooling system. With a sensible cooling system, the refrigerant is chilled and recirculated. Latent cooling may be accomplished with either a solid refrigerant, such as ice or solid carbon dioxide (dry ice), or a liquid. Ice and dry ice are not recommended as refrigerants for seed storage rooms.

Liquid Refrigerants

Mechanical refrigeration systems are based on the ability of liquids to absorb enormous quantities of heat as they vaporize. Vaporizing liquids provide a refrigeration system that can be easily controlled. Such systems can be started and stopped at will and the rate of cooling can be predetermined within narrow limits. The vaporizing temperature of the liquid can be regulated by controlling the pressure at which the liquid vaporizes. By using a closed system, the vapor can be readily condensed back into a liquid so that it can be used over and over again to provide a continuous flow of liquid for vaporization.

Since no one liquid refrigerant is best suited to all applications and operating conditions, the refrigerant selected should be the one best suited to meet the specific needs of each storage facility. Of all the fluids currently used as refrigerants, the one nearest the ideal general purpose refrigerant is dichlorodifluoromethane (CCl_2F_2). It is one of a group of refrigerants introduced under the trade name "Freon," but it is now manufactured under several other proprietary names. To avoid any confusion among trade names, this compound is now referred to as Refrigerant-12 or R–12.

Refrigerant-12 has a saturation temperature of −29.8° C, which means it can be stored as a liquid at ordinary temperatures only under pressure in heavy steel cylinders. Although an insulated space can be refrigerated by allowing liquid R–12 to vaporize in a container vented to the outside, this is not a practical method of refrigerating a seed storage room. The container in which the refrigerant vaporizes during refrigeration is called the evaporator and is an essential component of any mechanical refrigerating system.

Typical Mechanical Refrigeration System

A typical mechanical refrigeration or vapor-compression system consists of the following essential parts: (1) An evaporator to provide a heat transfer surface through which heat moves from the space being refrigerated into the vaporizing refrigerant; (2) a suction line to convey the refrigerant vapor from the evaporator to the compressor; (3) a compressor to heat and compress the vapor; (4) a hot gas or discharge line to carry the high-temperature, high-pressure vapor from the compressor to a condenser; (5) a condenser to provide a heat transfer surface through which heat passes from the hot gas to the condensing medium; (6) a receiving tank to hold the liquid refrigerant for future use; (7) a liquid line to carry the liquid refrigerant from the receiving tank to the refrigerant metering device; and (8) a refrigerant metering device to control the flow of liquid to the evaporator. The typical vapor-compression system is divided into a low and a high pressure side. The refrigerant metering device, evaporator, and suction line constitute the low part of the system; the compressor, discharge line, condenser, receiving tank, and liquid line constitute the high pressure side of the system.

Condensing Units

A condensing unit may be either air or water cooled. Condensing units of small horsepower are frequently equipped with hermetically sealed motor compressor assemblies. Large condensing units are usually water cooled.

System Capacity

The capacity of a refrigeration system is the rate at which it removes heat from the refrigerated space. Capacity is usually expressed in terms of units of heat removed per hour or its ice-melting equivalent. In other words, a mechanical refrigeration system that will cool at a rate equivalent to the melting of 1 ton of ice in 24 hours is said to have a capacity of 1 ton. The capacity of a mechanical refrigeration system depends on the weight of refrigerant circulated per unit of time and refrigerating effect of each pound circulated.

Compressor Capacity

The capacity of a compressor must be such that the vapor is drawn from the evaporator at the same rate at which it is produced. If the vapor is produced faster than the compressor can remove it, the accumulation of excess vapor will increase the pressure in the evaporator; this will, in turn, increase the boiling temperature of the refrigerant. If the capacity of the compressor is such that the vapor is removed too rapidly from the evaporator, the pressure in the evaporator will decrease and lower the boiling temperature of the refrigerant. In either case, the refrigeration systems will not function properly. For any good refrigeration system, the rate of vaporization must balance the rate of condensation of the vapor back to a liquid. Such a balanced system will function properly and refrigerate exceptionally well. For further information, see Dossat *(1961)*.

Controlling Humidity

Air is a mixture of gases, consisting primarily of nitrogen, oxygen, carbon dioxide, water vapor, and small percentages of rare gases. Each gas, including water vapor, exerts its own partial pressure in the mixture just as though the other gases were not present. The sum of these partial pressures equals the total pressure of the mixture. The amount of water vapor that can be contained in the air mixture is a constant value depending only on the temperature and pressure of the mixture. Thinking of humidity in terms of partial pressure makes it easier to understand the movement of moisture from one area to another, for moisture moves from a high to a low pressure area. It is therefore possible for moisture to move against the flow of air. Pressurizing a room will not prevent moisture from moving in, although it could slow down its entry.

Relative humidity is normally measured by taking dry-bulb and wet-bulb temperature readings and finding the intersection of those readings as plotted on a psychrometric chart. The point of intersection will correspond to a particular relative humidity. Relative humidity can be changed by raising the air temperature without changing absolute humidity. For further information, see Dryomatic Division, LogEtronics *(1965).*

Moisture Movement Between Air and Materials

Structural materials, such as wood and cement, contain moisture throughout, whereas such materials as steel and glass hold moisture on their surface. The rate of moisture vapor movement between such materials and air is determined by the difference in moisture vapor pressure between them. If their moisture vapor pressures are equal, moisture will not move from one to the other.

When moist seeds are placed in a dry atmosphere, moisture will flow from the seeds into the atmosphere. Because the air cannot hold nearly all the moisture held in the seeds, the air will soon become saturated with the moisture given off by the seeds, and unless new dry air is provided, drying of the seeds will stop. Because construction materials contain moisture, that moisture as well as the moisture in the seeds has to be removed. Once the room and the seeds reach moisture equilibrium with the desired relative humidity, the drying system has only to remove the moisture that enters the controlled atmosphere room through door openings, leakage through seams and cracks, and penetration of the barrier material. The storage area is kept at a designated relative humidity, which, in turn, prevents any change in the moisture content of stored seeds once they have reached moisture equilibrium with the maintained relative humidity. For additional information, see Hass *(1961, 1965),* Sijbring *(1963),* Dryomatic Division, LogEtronics *(1965),* and Munford *(1965).*

Refrigeration-Type Humidity Control Systems

The refrigeration-type dehumidifier draws warm, moist air over a metal coil with fins spaced far enough apart to permit partial frosting and still allow for sufficient passage of air. The frost can be removed by electrical heater, hot gas, or water defrosting at regular intervals regulated by a timeclock. To be effective at low temperatures, a refrigeration-type dehumidification system must cool the air below the desired temperature and reheat it to the desired temperature. Air-handling units are available with built-in refrigeration coils, electric defrosters, and reheat coils. An engineered unit such as described may be better than a piecemeal assembly because the correct wattage is provided in the unit.

The simplest method of regulating the reheat in a room is to wire the reheat coils through a relay to a humidistat. Although the refrigeration-reheat system can effectively control relative humidity as well as temperature, it is not the only system available and in some cases may not be the best. When very low relative humidity is to be maintained, operation of the refrigeration-reheat system becomes very expensive.

Desiccant-Type Humidity Control Systems

Dehumidifiers using liquid or solid desiccants in conjunction with refrigeration can frequently reduce the cost of maintaining very low relative humidities. These desiccants absorb moisture vapor from the airstream and later eject it outside the room. Desiccant dehumidifiers generally use dry chemicals for small systems and salt solutions for systems with extremely large volumes of air. For seed storage facilities, dry desiccant systems are almost invariably used. The dehumidifier incorporates one or two beds of granulated silica gel or activated alumina, which can absorb much water vapor. For example, silica gel can absorb as much as 40 percent of its weight in water vapor at 100-percent relative humidity and proportionally less at lower relative humidity.

With the two-bed system most frequently used for seed storage facilities, the air is circulated through one bed at a time. One bed is recharged, or dried out, while the other is taking up moisture. The switch from one bed to the other is usually programed by a timeclock for maximum efficiency.

Recent developments receiving wide application are the rotary drum or cylinder (fig. 31) and rotary disc (fig. 32) dehumidifiers. The rotary dehumidifiers have one or more beds divided into two airstreams by sealing strips. The bed or beds rotate slowly, and while part of each bed is absorbing water vapor from the airstream, the remainder is being recharged. The end result is much the same as for the two separate bed systems except that the rotary systems seem to have a higher drying

PN–5410, PN–5411

FIGURE 31.—*Above*, rotary cylinder dehumidifier used to control relative humidity in seed storage rooms; *below*, arrangement of cylinder, reactivation heaters, and seals in this dehumidifier. (Courtesy of Dryomatic Div.)

capacity for a given volume of air passing through the dehumidifier. In contrast to the refrigeration type, desiccant dehumidifers are not affected by the freezing of moisture and can handle air at −40° to +80° C as long as the appropriate desiccant is used.

PN–5412, PN–5413

FIGURE 32.—*Above*, rotary disc dehumidifier used to control relative humidity in seed storage rooms; *below*, arrangement of discs and seals in this dehumidifier. (Courtesy of Dryomatic Div.)

For seed storage facilities where both a low temperature and a low relative humidity are needed, a combination refrigeration-desiccant dehumidifier system will probably provide the most reliable temperature and relative humidity control at the least cost. Maintenance cost of dehumidification systems is usually small. For desiccant systems, the desiccant material should be changed every 3 to 5 years. For additional information, see Hass *(1961, 1965)*, James *(1962)*, Cooke *(1966)*, Beck *(1966, 1972)*, and Welch *(1967)*.

Common-Sense Practices

Ventilation

Seeds stored wet tend to get hot whether stored in bulk or in bags unless heat is rapidly dissipated as it is generated. Both heat and excess moisture can be dissipated through ventilation. A steady stream of air moving through a bin of loose seeds or a warehouse full of bags of seeds will collect the heat as it is generated and transport it away. Preventing heat buildup is essential to maintaining good germination. A good ventilating system will soon pay for itself through improved seed quality.

Stacking Seed Bags

Most processed seeds are handled in bags, which are stored in stacks. Bags of seeds of each genus, species, and cultivar are stored in separate stacks. To allow for proper ventilation, stacks of seed bags should be well spaced, both between the bags and between the stacks. Bags of seeds must be stacked carefully to prevent slippage and falling. Falling bags are hazardous to employees as well as subject to breakage on impact. Stacks that are excessively high often cause the bottom bags to burst. It is especially important that bags moved by forklift be stacked securely on pallets.

Removing Seeds From Controlled Storage

Dry seeds removed from cold storage and exposed to a warm, humid atmosphere will absorb moisture readily. Unless preventive steps are taken, bags of cold seeds will become moist from the condensing atmospheric moisture. Moisture condensation can be prevented by warming the seeds in a dry atmosphere or by warming in a room ventilated with rapidly moving air. Use of moisture-barrier containers can prevent condensed moisture from coming into contact with the seeds. Proper aeration can bring cold bulk seeds up to normal atmospheric temperature without surface accumulation of condensed atmospheric moisture. Unless seeds held under controlled atmosphere storage are kept dry after removal from storage, the protective effects of controlled storage will soon be lost.

Warehouse Cleanliness

Warehouse cleanliness, or lack of it, can markedly affect seed quality. A clean warehouse has fewer rodents and insects and they are more easily controlled. A clean warehouse reduces the chances of accidental mixtures and personnel accidents. It also presents a better company image to visitors.

PACKAGING AND PACKAGING MATERIALS

Requirements of Different Situations

Seed-packaging methods are many and varied (Bass et al., *1961*). The equipment for filling packages ranges from a simple spoon or seed scoop, to the manually controlled gravity flow from a bin, to the high-speed automatically controlled small packet filler. Package-filling equipment has no effect on the genetic or physiological quality of seeds; however, it may affect the physical quality of seeds through impact or undue pressure. Heavy seeds, especially bean, corn, pea, and soybean, can be fractured by impact on the individual seed by either striking or being struck by a hard object or firm surface. Seeds containing too much or too little moisture damage easily on impact. Damage from repeated impact is accumulative. Seeds may also be injured by force feeding through a restricted opening or between pressure rolls. Physical damage can occur during any handling operation from harvest to planting.

Types of Packages

Packages for processed seeds may be burlap, cotton, paper, or film (plastic, foil) bags, metal or fiberboard cans or drums, glass jars, fiberboard boxes, or containers made of various combinations of materials. The types and sizes of containers used for wholesale distribution of field seeds are usually the same as those used for retail sales. For vegetable and flower seeds, wholesale containers may be large fabric or multiwall paper bags, large fiberboard or metal cans or drums, or fiberboard boxes, whereas retail containers are usually small paper, film, or film-laminated envelopes, small fiberboard boxes, or small metal cans.

The packaging materials, methods, and equipment used are dictated by the kinds and amounts of seeds to be packaged, the type of package, duration of storage, storage temperature, relative humidity of the storage area, whether packaging is for wholesale, retail, or local use, and geographical area where the packaged seeds will be stored, exhibited, or sold. Packages that will contain seeds and protect most physical qualities of seed lots are made of materials with sufficient tensile strength, bursting strength, and tearing resistance to withstand normal handling. However, such materials do not protect seeds against either insects and rodents or changes in moisture content unless special protective qualities are built into them. Researchers frequently use whatever container happens to be convenient without regard to the moisture protection it may or may not afford. However, there is an increasing awareness of the savings in time and expense that are realized by using suitable moisture-barrier containers for storing valuable breeding stocks.

Seeds stored for a short time or under cold, dry conditions will retain good viability in porous paper or fabric containers, whereas seeds stored or marketed under tropical conditions (Ching, *1959*) will lose viability rapidly without maximum moisture protection. Except in tropical and subtropical climates, many kinds of seeds do not require special moisture protection during the first winter after harvest when held in the area where produced. However, seeds carried over to the second planting season often require drying and packaging in moisture-barrier containers to prevent loss of viability and vigor.

How Packages Are Filled

Except in very limited operations that employ hand-filling, seeds to be packaged are delivered to hopper bins above automatic or semiautomatic filling machines. Seeds may come to the hopper from bulk storage bins by gravity flow through pipes and by airlift, belt conveyors, forklifts, or the human shoulder. Since practically all seeds are sold by weight, it is necessary to put a predetermined amount into the individual packages. Even seeds that appear to be sold by volume are associated with weight; namely, a bushel of corn equals 56 pounds, a bushel of wrinkled peas 56 pounds, and a bushel of smooth seeded peas 60 pounds. Most package-filling equipment has built into it a seed-measuring device, or it is manually or automatically controlled by a signal from a weighing device.

How Containers Are Presented to the Filler

Large nonrigid containers of burlap, cotton, lined bags, multiwall and 7- and 10-mil polyethylene bags are usually presented to the filler manually. They are held in place by hooks, clamps, or by hand during filling. Polyethylene and similar materials may be formed into bags from sheets or rolls, filled, and sealed in a continuous operation. Large rigid containers, other than metal or glass, are usually formed into an open box by hand and presented to the filling equipment either manually or by a conveyor that automatically positions each container. Small, semirigid, preformed containers may be opened with a jet of air and automatically positioned for filling. Rigid containers come from the manufacturer ready for filling. They are positioned for filling either manually or by conveyor.

Placing seeds in rolls of tape may be considered a form of packaging.

During the midtwenties, equipment was developed for forming the tape, placing the seeds, and making the roll. Although this form of packaging did not become popular at that time, there currently appears to be a revival of interest in this method.

Weighing and Measuring Devices

Scales for weighing seeds range from the large truck scale, which weighs an entire load at a time, to the common beam scale or an elaborate scale that activates either a pneumatic or electric device that shuts off the flow of seeds when a predetermined weight or volume is reached. In retail stores many types of scales are used, some of which automatically compute the cost to the customer based on the price per pound. Some scales are so delicate they can weigh a few small seeds, such as petunia. Measuring devices range from a simple spoon or scoop, to a calibrated cylinder or cup, to the measuring device associated with the automatic scales.

How Containers Are Closed

Hand-tying of cotton and burlap bags has been largely replaced by sewing. Although there is still some sewing by hand, most of it is done by machine. Polyethylene and other thermoplastic materials are usually sealed by applying heat at 93.3° to 204.4° C to the film for a given time while the film is under pressure. Within limits each kind and thickness of material has specific temperature, time, and pressure requirements (table 23) for proper sealing.

TABLE 23.—*Approximate heat-sealing requirements for selected thermoplastic materials*[1]

Material	Heat sealing	Time	Pressure
Film	°C	*Seconds*	*Lb/in²*
Polyethylene (PE) density:			
Low	[2]120–205	---	---
Medium	150–205	---	---
High	120–220	---	---
Polyester, 0.25 mil	150–205	0.2–2	20–60
Metalized polyester, 0.2 mil	135–205	.2–2	20–60
Laminate			
Scrim/PE/Foil/PE	275	±3	40–60
Paper/Saran/PE	150	1	40
50–lb kraft/PE/ 0.05 mil foil/PE	190	±7	40–60

[1] Specific temperature, time, and pressure vary with kind and thickness of each film and kinds of films in each laminate.

[2] Heat sealing varies with thickness of material.

Heat sealers range from small hand-operated rollers or bars, to foot-operated bars, jaws, and clamps, to elaborate automatic bag- or pouch-forming, filling, and sealing machines. Some sealers use thermostatically controlled bars, bands, or rollers, and others use high-intensity thermal impulses of short duration. Most are readily adjustable for use with many kinds and thicknesses of material. However, the quality of the seal produced by hand- and foot-operated equipment depends on the skill of the operator. Properly adjusted automatic sealers seldom produce a poor seal. Considerable experience is required before an operator can consistently apply the appropriate pressure for the precise dwell time required to produce a good seal. It is especially important that seed packages of proper moisture-barrier materials be well sealed as a leak in the seal will soon negate the moisture protection provided by the material.

Semirigid or rigid containers, other than metal or glass, are usually sealed with cold or hot glue applied either by hand or automatically by machine. Most machines that apply glue also form and fold the open ends of the containers and place them under pressure until the glue has set. Rigid containers, such as fiber drums, may have slip-on caps or lids that clamp into position. These lids are applied manually, whereas lids for metal cans are applied mechanically. Can sealers are manually operated, semiautomatic, or fully automatic. Usually the equipment is the same as that used by food processors for hermetically sealed cans. A partial vacuum or a gas can be introduced into the can when semiautomatic and fully automatic equipment is used for sealing. Some cans are sealed with pressure lids similar to those used on paint cans.

Packaging Field Seeds

Field seeds are generally packaged in burlap, osnaburg, or seamless and multiwall paper bags containing 50 or 100 pounds or one-half to 3 bushels of seeds. A few companies use moisture-barrier packages, such as elastic multiwall paper bags with either an asphalt or a polyethylene ply in the multiwall, burlap or cotton bags with polyethylene liners, and burlap/asphalt/paper-laminated bags for cereal, soybean, and hybrid sorghum seeds. Many hybrid corn seed producers either package all or part of their crop each year in moisture-resistant packages or hold their seed in controlled storage. Moisture-resistant bags used for seed corn are made of multiwall/asphalt/paper, elastic multiwall/asphalt/paper, multiwall and elastic multiwall/paper with a polyethylene-barrier ply, or 7- or 10-mil polyethylene.

A valve-type polyethylene bag has been developed that prevents any loss of material while filling and that seals easier than the conventional bag. The valve-type bag also permits easy introduction of fumigants or gases into the filled bag prior to sealing.

Some companies are packaging hybrid corn in acre units, each package containing the exact quantity of seed required to plant a specified number of acres.

Turf grass seed for the wholesale market is usually put in fabric bags of 25- to 100-pound capacity. Sometimes a 4- or 5-mil polyethylene liner is used for moisture protection. At retail, turf grasses are sold as single species or combinations of species packaged in paper, cloth, or polyethylene bags, metal cans, or cardboard boxes. The cardboard boxes may be plain, polyethylene, or wax paper lined, foil or wax coated, or wax paper or foil overwrapped.

Tobacco seeds are packaged in paper packets, cardboard boxes or cylinders, or metal cans of ½- or 1-ounce capacity.

Moisture-resistant containers, such as wax- or polyethylene-coated cardboard boxes or drums and polyethylene, cotton, or paper combinations, are being used for various forage grasses and legumes. Cottonseeds previously dried to 6- to 8-percent moisture content are packaged in burlap/asphalt/paper bags or wax- or polyethylene-coated boxes. Large wood or steel pallet boxes with covers are being used extensively in the seed industry for bulk storage of both uncleaned and cleaned seeds. Such bulk storage reduces handling costs and increases the efficiency of cleaning plant operations by permitting a greater amount of mechanization. There is no handwork involved in handling pallet boxes, but with bags or small boxes, a great deal of handwork is required.

Packaging Vegetable and Flower Seeds

In the vegetable seed industry, the packaging method is determined primarily by the type of customer for which the seeds are packaged. Seeds for wholesale distribution are usually packaged in fabric bags of 25- to 100-pound capacity. The bags may or may not have polyethylene liners. Beans, peas, and sweet corn are usually packaged in multiwall/asphalt/paper bags containing 25 to 100 pounds of seed. For retail sales, most vegetable seeds are packaged in paper packets, sometimes with foil inserts, containing a fraction of an ounce to several ounces; cardboard boxes holding a few ounces; and 2- to 4-mil polyethylene, cellophane, acetate, paper, and foil-laminated bags with capacities of 1 to 5 pounds. Various kinds of vegetable seeds are also packaged in hermetically sealed metal cans of various sizes. These containers are especially beneficial for shipments overseas and into tropical areas.

Flower seeds at the wholesale level are packaged in paint-type cans, glass jars, and fabric bags with polyethylene liners. For retail sales, paper, paper/polyethylene, acetate, cellophane, and foil-laminated packets are the predominate types. Wholesale packages of flower seeds usually contain several pounds, whereas retail packets contain a given number of seeds or a fraction of an ounce.

Package Labeling

Researchers and seedsmen alike need to identify the contents of each container, both as to species and cultivar. Frequently other information, such as percentage of live seeds, purity, noxious weed seed content, and seed treatment, if any, must also be recorded as required by law. This may be done by printing the required information on a tag attached to flexible cotton or fiber bags, by printing on a label glued to tin cans, cardboard boxes, or cardboard or metal drums, or by printing or stamping the information directly on the container, such as lithographing cans or embossing metal lids.

Porous Packaging Materials

Burlap or hessian bags are made of good quality jute yarn in a variety of fabric constructions. Since burlap is exceptionally strong, burlap bags lend themselves to stacking high in storage and to rough handling, and they can be reused many times. Seed bags of cotton are made of sheeting, printcloth, drill, and osnaburg fabrics as well as a special seamless material. Osnaburg and seamless fabrics have the greatest strength and tear resistance of the cotton materials. Bags made of these materials are reused many times, whereas bags of other cotton materials are used only once. Reuse of fabric bags is largely confined to storage of unprocessed seed as certification standards require new bags for shipment of processed seeds.

Paper products are extensively used for seed packaging. Small seed packets are mostly made of bleached sulfite or bleached kraft paper and surface coated with a very white clay to facilitate printing. Basically paper packets are designed to contain, without loss, a given quantity of seeds, not to protect seed viability.

Many paper seed bags are of multiwall construction consisting of several layers of smooth or crinkled paper. Multiwall paper bags are produced in a variety of constructions, each designed for a specific purpose. Ordinary multiwall bags have poor bursting strength and consequently when piled high, the bottom bags burst. The top bags in high piles often slip. Under very dry conditions, multiwall bags dry out and become brittle along folds and at wear points. Elastic multiwall bags consist of several layers of crinkled paper. Elastic materials cannot be evaluated by the usual physical test data for tensile and tear strength because the entire principle of the elastic multiwall bag depends on built-in stretchability.

Cardboard, in the form of boxes and cans, is used extensively in seed packaging. Cardboard containers protect most physical qualities of seed and are well adapted for automatic filling and sealing equipment. Porous packaging materials adequately contain or hold the seeds and protect them from mechanical mixtures, but they do not provide moisture protection.

Moistureproof Materials

Metal containers when properly sealed provide an absolute moisture and gas barrier and completely shield the products from any effects of light. Metal containers provide complete protection against rodents, insects, changing humidity, flood, and harmful fumes. Metal cans are well adapted to high-speed automatic filling and sealing.

Glass containers are not widely used in seed packaging. Although glass provides essentially the same protection as metal, its fragility makes it less desirable for commercial packaging. Glass containers are frequently used in research and as display receptacles in seed and hardware stores, where bulk sales of seed are made. Glass containers are often used by home gardeners for carryover of small quantities of seed from one season to the next.

Adequate drying before sealing is absolutely essential for safe storage of seeds in airtight, moistureproof containers, especially for seed stored at warm temperatures or shipped to tropical areas. Much research has been done on the effects of sealed moistureproof storage on seed longevity. Some workers used air-dried seeds, whereas others carefully dried the seeds before sealing.

Moistureproof Storage

Response of Different Crop Seeds to Moistureproof Storage

Cereal Seeds.—Numerous studies have shown that rice seeds held in airtight storage outlive seeds held in open storage provided they are adequately dried before they are placed in airtight storage (Vibar and Rodrigo, *1929*; Kondo and Okamura, *1930, 1932–33, 1938*; Rodrigo, *1935, 1953*; Kondo and Terasaka, *1936*). Rice seeds dried by high heat before sealing were dead after 26 to 28 years, but seeds dried by the sun to 11- to 13-percent moisture were stored safely for 30 years (Kondo and Okamura, *1932–33*). Cultivar differences in storability were demonstrated by Vibar and Rodrigo *(1929)*, who reported that the germination of seeds of the rice cultivar Hambas declined 18 percent in 51 months, whereas seeds of the Inintew cultivar declined only 5 percent.

Because corn seed is usually planted the year after harvest, people in the Corn Belt usually have no problem with storage. However, corn is now grown in areas where temperature and humidity conditions frequently cause rapid loss of viability during storage. Research has shown that sealed storage prevents rapid loss of viability of corn seeds provided their moisture content is sufficiently low when the seeds are sealed (Vibar and Rodrigo, *1929*; Rodrigo, *1935, 1953*; Kaihara, *1951*; Barton, *1960b*). Sealed seeds with 11-percent moisture content maintained full viability for 9 years at −5° C but showed reduced viability after 1 year at 30° (Barton, *1960b*). Although sealed glass jars were used in this study, similar results could be expected with any type of sealed moistureproof container, such as a metal can or drum or a multiwall bag with a foil layer.

Sorghum seeds in commerce usually are not stored longer than over winter. However, plant breeders frequently need to hold seeds for many years. Sorghum seeds dried and sealed in glass bottles retained their viability longer than did similar seeds stored in gunny sacks (Krishnaswamy, *1952*). We found that cultivar RS619 sorghum seeds retained their germination up to 8 years at −12° and −1° C, whether in paper envelopes or sealed metal cans (table 24). Sealed seeds at 10° retained viability significantly better than did unsealed seeds. At higher temperatures the results were variable. Some of the variability can be attributed to differences in seed moisture content. The seeds in sealed metal cans had an initial moisture content that did not change with time. The moisture content of the seeds in paper envelopes adjusted with time to moisture equilibrium with the relative humidity in the storage chamber. The low relative humidity at the higher temperatures allowed the seeds to attain a low moisture content and partly offset the effects of temperature on longevity.

TABLE 24.—*Germination of 5 kinds of seeds at 3 initial moisture contents after storage in paper envelopes or sealed metal cans under 5 temperature and relative humidity conditions for 4 and 8 years*

Storage temperature (C), relative humidity (RH) (percent), and initial seed moisture content (percent)	Germination[1] after storage for indicated container and years				
	Paper envelope			Sealed metal can	
	0	4	8	4	8
	Percent	*Percent*	*Percent*	*Percent*	*Percent*
			CRIMSON CLOVER		
−12° and 70 RH:					
4	82	79	78	84	84
7	92	84	78	87	87
10	91	83	83	87	80
−1° and 60 RH:					
4	82	80	66	80	85
7	92	78	71	89	83
10	91	85	75	86	70

See footnote at end of table.

TABLE 24.—*Germination of 5 kinds of seeds at 3 initial moisture contents after storage in paper envelopes or sealed metal cans under 5 temperature and relative humidity conditions for 4 and 8 years*—Continued

Storage temperature (C), relative humidity (RH) (percent), and initial seed moisture content (percent)	Germination[1] after storage for indicated container and years				
	Paper envelope			Sealed metal can	
	0	4	8	4	8
	Percent	*Percent*	*Percent*	*Percent*	*Percent*
			CRIMSON CLOVER—con.		
10° and 60 RH:					
4	82	74	44	82	82
7	92	71	41	80	76
10	91	77	45	75	42
21° and 30 RH:					
4	82	67	33	83	80
7	92	71	34	83	81
10	91	76	34	43	5
32° and 15 RH:					
4	82	83	71	75	80
7	92	76	69	76	69
10	91	78	73	3	0
			LETTUCE		
−12° and 70 RH:					
4	97	95	92	96	96
7	97	95	96	95	98
10	95	95	95	93	94
−1° and 60 RH:					
4	97	91	84	94	90
7	97	95	90	97	76
10	95	96	96	94	91
10° and 60 RH:					
4	97	94	0	93	90
7	97	94	0	94	1
10	95	92	0	21	0
21° and 30 RH:					
4	97	11	0	89	90
7	97	6	0	4	0
10	95	14	0	0	0
32° and 15 RH:					
4	97	89	2	90	36
7	97	84	2	0	0
10	95	93	0	0	0
			SAFFLOWER		
−12° and 70 RH:					
4	95	95	86	94	87
7	94	94	88	94	83
10	95	95	88	91	79

See footnote at end of table.

TABLE 24.—*Germination of 5 kinds of seeds at 3 initial moisture contents after storage in paper envelopes or sealed metal cans under 5 temperature and relative humidity conditions for 4 and 8 years*—Continued

Storage temperature (C), relative humidity (RH) (percent), and initial seed moisture content (percent)	Germination[1] after storage for indicated container and years				
	Paper envelope			Sealed metal can	
	0	4	8	4	8
	Percent	*Percent*	*Percent*	*Percent*	*Percent*
SAFFLOWER—con.					
−1° and 60 RH:					
4	95	93	90	97	91
7	94	95	86	90	90
10	95	94	88	92	72
10° and 60 RH:					
4	95	96	81	95	89
7	94	96	88	91	85
10	95	92	86	0	0
21° and 30 RH:					
4	95	90	64	94	90
7	94	89	61	0	0
10	95	89	54	0	0
32° and 15 RH:					
4	95	63	53	93	89
7	94	63	60	0	0
10	95	65	53	0	0
SESAME					
−12° and 70 RH:					
4	94	95	92	94	92
7	92	92	93	90	89
10	88	93	89	0	0
−1° and 60 RH:					
4	94	93	94	91	92
7	92	92	90	89	87
10	88	90	90	6	0
10° and 60 RH:					
4	94	96	88	92	90
7	92	92	95	68	0
10	88	88	84	0	0
21° and 30 RH:					
4	94	91	88	93	91
7	92	90	88	0	0
10	88	91	83	0	0
32° and 15 RH:					
4	94	94	86	95	87
7	92	95	88	0	0
10	88	95	70	0	0

See footnote at end of table.

TABLE 24.—*Germination of 5 kinds of seeds at 3 initial moisture contents after storage in paper envelopes or sealed metal cans under 5 temperature and relative humidity conditions for 4 and 8 years*—Continued

Storage temperature (C), relative humidity (RH) (percent), and initial seed moisture content (percent)	Germination[1] after storage for indicated container and years				
	Paper envelope			Sealed metal can	
	0	4	8	4	8
	Percent	*Percent*	*Percent*	*Percent*	*Percent*
			SORGHUM		
−12° and 70 RH:					
4	92	96	94	93	90
7	95	93	95	94	94
10	91	93	92	97	94
−1° and 60 RH:					
4	92	93	95	91	92
7	95	94	89	94	94
10	91	92	92	92	87
10° and 60 RH:					
4	92	90	71	90	90
7	95	91	76	93	90
10	91	91	76	93	92
21° and 30 RH:					
4	92	85	55	94	83
7	95	84	84	92	86
10	91	80	80	84	72
32° and 15 RH:					
4	92	77	50	90	65
7	95	76	46	85	64
10	91	73	40	69	0

[1] Least significant difference at 5-percent level of probability is 10 percent.

Forage Grasses.—Seeds of *Echinochloa, Eleusine, Panicum, Pennisetum, Paspalum,* and *Setaria* retained their viability longer when dried and stored in sealed bottles than when stored in gunny sacks. Sealed *Echinochloa, Panicum,* and *Setaria* seeds retained approximately 70-percent germination for 38 months (Krishnaswamy, *1952*). Crested wheatgrass, intermediate wheatgrass, and smooth bromegrass seeds retained high viability best at −18° C whether sealed or open (Knowles, *1967*). At 1°, seeds stored in sealed glass jars held their germination better than those in open storage. At 21°, seeds in sealed containers retained a higher percent viability than did seeds in paper envelopes; however, the viability of sealed seeds sharply declined during 4 years of storage.

Turf Grasses.—Perennial ryegrass seeds containing 6-, 8-, 12-, 16-, and 20-percent moisture were stored in sealed metal cans at 3°, 22°, and 38° C in a warehouse in Oregon (Ching et al., *1959*). After 3 years of storage, the 6-percent moisture seeds showed some decline in germination at 38°, seeds with 8-percent moisture retained good germination at all temperatures except 38°, but seeds with 12- and 16-percent moisture remained viable only at 3°. Seeds with 16- and 20-percent moisture deteriorated within 3 months when stored at 22° and 38°. Ching and Calhoun *(1968)* reported that after 10 years of sealed storage, 6-percent moisture ryegrass seeds stored at 22° and 38° gave significantly lower laboratory germination percentages than new seeds and similar seeds held at other temperatures; however, field emergence was not significantly different from that of all other seed lots.

Kentucky bluegrass seeds with 8.7-, 6.2-, and 4.9-percent moisture packaged in sealed tin cans and stored at 1°, 10°, and 21° C germinated at approximately their original level after 30 months in storage regardless of the seed moisture level (Bass, *1960*). Creeping red fescue containing 11.4-percent moisture in sealed metal cans dropped sharply in viability during the first 3 months of storage at all temperatures; seeds sealed with lower moisture levels retained good viability longer, especially at the lower temperatures.

Forage Legumes.—Crimson clover seeds containing 6-, 8-, 12-, 16-, and 20-percent moisture were stored in sealed metal cans at 3°, 22°, and 38° C in a warehouse. Seeds with 16- and 20-percent moisture at 22° and 38° deteriorated within 3 months, those with 8-percent moisture were preserved well except at 38°, and those with 12- and 16-percent moisture remained viable only at 3°. The 6-percent moisture crimson clover seeds remained highly viable after 3 years of storage under all temperature conditions. (Ching et al., *1959*) Ching and Calhoun *(1968)* reported that after 10 years of sealed storage only 16-percent moisture crimson clover seeds held at 3° germinated significantly lower than fresh seeds in laboratory tests. In the field there were no significant differences in emergence between fresh and 10-year-old seeds regardless of the temperature under which the sealed seeds had been stored.

In our studies, crimson clover seeds with 4- and 7-percent moisture retained higher viability or germination in sealed metal cans than in paper envelopes at all storage conditions (table 24). However, the 10-percent moisture seeds lost viability more rapidly in the cans than in the paper envelopes when stored at 21° and 32° C. This was, of course, because the seeds in paper envelopes lost moisture during the early weeks of storage, whereas those in the cans did not. The 10-percent moisture seeds in metal cans actually declined 50 percent in viability during the first year of storage at 32°, whereas the seeds in paper envelopes declined only 11 percent.

These data show that seeds in hermetically sealed containers may lose viability more rapidly than those in open storage if seed moisture content, storage temperature, or both is too high.

Alfalfa seeds with 6.8-percent moisture germinated 78 percent after 24 years in sealed glass bottles, and red clover seeds with 6.9- and 6.7-percent moisture germinated 74 and 71 percent, respectively (Nutile, *1958*).

Fiber and Oil Crops.—Flaxseeds containing 7- to 8-percent moisture when stored in metal containers retained good germination for 13 to 16 years at temperatures prevailing in Mandan, N. Dak. (Dillman and Toole, *1937*). Cottonseeds with 6- to 8-percent moisture in sealed glass jars kept for 7 to 10 years at 21° C without loss of viability (Simpson, *1942, 1953*). Good quality cottonseeds with 7- to 11-percent moisture content retained high viability for over 25 years in sealed metal cans at 1° (Pate and Duncan, *1964*).

Hempseeds with 8.6-percent moisture or less retained good germination for 3½ years at 2° C (Crocioni, *1950*). Seeds with 5.7-percent moisture still retained their initial viability after 15 years of sealed storage at −10° and 10° (Toole et al., *1960*; Clark et al., *1963*). Two lots of seeds with 6.2-percent moisture in sealed cans retained full viability for 12 years at −10°, 0°, and 10°, whereas a third lot of lower initial germination decreased 23 percent in germination during 12 years of storage. The two high germinating lots of hempseeds when stored with 9.5-percent moisture retained good viability at −10° and 0° but deteriorated rapidly at 10°. Seeds of the poorer germinating lot declined 10 percent at −10°, 27 percent at 0°, and lost all viability at 10° during 12 years of storage.

Kenaf seeds with 8-percent moisture retained their initial viability for 12 years when stored sealed at −10°, 0°, and 10° C. Seeds with 12-percent moisture, which had retained full viability for 5½ years at −10° and 0° and showed a significant loss of viability after 4½ years at 10° (Toole et al., *1960*), were nonviable after 12 years at 10° and had declined sharply at 0° (Clark et al., *1963*).

For both safflower and sesame seeds, careful control of seed moisture is absolutely essential for sealed storage. Safflower seeds containing 4- and 7-percent moisture retained their germination about equally well in paper envelopes and in sealed metal cans when stored for 8 years at −12°, −1°, and 10° C (table 24). The 4-percent moisture seeds in sealed cans retained better viability than did the envelope-stored seeds at 21° and 32°. However, 7-percent moisture seeds in sealed metal cans lost all viability in less than 4 years at 21° and in less than 1 year at 32°. Ten-percent moisture seeds in sealed metal cans retained fair viability for 8 years at −12° and −1°; however, all were dead in less than 3 years at 10° and in less than 1 year at 21° and 32°. The seeds in paper envelopes at 21° and 32° markedly decreased in viability, but no sample was completely dead regardless of the initial moisture content or storage temperature. For safe, sealed storage of safflower seeds at a temperature greater than 10°, seed moisture content must be reduced to 4 percent or less before sealing.

Sesame seeds in sealed metal cans did not retain satisfactory germination when seed moisture content exceeded 7 percent for storage at −12° and −1° C (table 24). At 10° and higher, seed moisture content must not exceed 4 percent. Ten-percent moisture seeds in sealed metal cans do not store well at any temperature. The 10-percent moisture seeds at 10° germinated 73 percent after 1 year but were nearly all dead at the end of the second year. Seeds at 21° and 32° lost all viability in less than a year. As with safflower, sesame seeds are very sensitive to both seed moisture content and storage temperature and must be carefully dried to 4-percent moisture before sealing unless the seeds are to be held continuously at −12° or −1°.

Under Philippine conditions the germination of air-dry soybean seeds stored in sealed containers lost viability after 23 months. (Vibar and Rodrigo, *1929*) and 54 months (Rodrigo, *1953*). In Illinois the germination of air-dry soybeans in sealed containers dropped only 2 percent during the first year but declined sharply thereafter, and nearly all seeds were dead within 8 years (Burlison et al., *1940*). The germination of 'Illsoy' at 7 years was comparable to 'Manchu' at 4 years and 'Lexington' at 5 years, indicating cultivar differences in keeping quality (Burlison et al., *1940*).

Soybean seeds conditioned to various moisture contents, then sealed in pint jars, and stored at various temperatures lost viability at a different rate at each storage condition (Toole and Toole, *1946*). Cultivars of Mammoth Yellow and Otootan gave essentially the same results. Seeds with 18-percent moisture at 30° C were nonviable in less than 3 months, and similar seeds at 10° died within 2 years. Seeds stored at subfreezing temperatures retained good germination for 6 years but were unsatisfactory for planting after 10 years. Air-dry seeds of 13.5-percent moisture, however, retained essentially their full initial viability for 10 years at 2° and −10°. Soybean seeds dried to 9-percent moisture or less before sealing retained full viability for 10 years at 10°, 2°, and −10° (Toole and Toole, *1946*).

Peanuts in sealed storage retained satisfactory germination up to 37 months (Vibar and Rodrigo, *1929*; Rodrigo, *1953*).

Vegetable Seeds.—Onion seeds kept very well in sealed storage (Brison, *1941, 1942*). Seeds dried to 6.4-percent moisture before sealing germinated 90 percent after 13 years at room temperature (Brown, *1939*). However, in another study (Asgrow Seed Co., *1954*), 6.3-percent moisture content seeds retained good viability for only 3 years at 32° C.

Although onion seeds dried to below 8-percent moisture before sealing kept well for 4 years at 5° (Beattie and Boswell, *1939*), 6.7-percent moisture is recommended for longer storage. Dry seeds in sealed containers in a laboratory lost viability in 38 months (Rodrigo, *1953*). Sealed onion seeds held in cold storage at 5° to 10° without deterioration for 2 years did not deteriorate when held 3 months at natural conditions after removal from storage (Myers, *1942*). This indicates that adequately dried onion seeds held sealed in cold storage can be safely marketed after removal from cold storage.

Rodrigo (*1935, 1953*) reported that under Philippine conditions the viability of air-dried mung beans stored in sealed containers in a laboratory was maintained up to 201 months. Cowpeas similarly stored remained viable for 123 months. White-seeded tapilan beans lost viability gradually over 10 years, whereas black-seeded tapilan beans maintained their original viability (Vibar and Rodrigo, *1929*; Rodrigo, *1935*). Yellow tapilan seeds lost viability in 138 months, whereas black tapilan seeds maintained viability up to 201 months (Rodrigo, *1953*).

Seeds of 'Buff Cross,' 'Mahogany Brown,' and 'Will's Red Kidney' beans germinated 20 percent, whereas 'Improved Michigan Robust' seeds germinated 50 percent after 38 years of storage in sealed canning jars in a laboratory at Geneva, N.Y. (Waters, *1962*). For 'Top Notch Golden Wax' and 'Bountiful' bean seeds stored open and sealed at −18°, −2°, 5°, 10°, 20°, and 30° C and ambient temperature in a laboratory at Yonkers, N.Y., sealing extended longevity when storage was in a humid room, even as low as 5° (Barton, *1966a*). Sealing was without effect at −18° and −2° for up to 15 years. Storage at 30° caused rapid deterioration in both open and sealed containers.

The germination and vigor of canned cucumber seeds did not change significantly in 36 months when seed moisture was 5.3 percent or less at 32° C or lower (Asgrow Seed Co., *1954*). With higher seed moistures, temperature becomes more critical. Seeds with 9.4-percent moisture became worthless in 3 months at 32° but held viability and vigor well for 30 months at 15.6°. Tin cans were superior to chlorinated rubber, waxed cellophane, and paper bags inside linen ones (Coleman and Peel, *1952*). Dry seeds in sealed containers in a laboratory lost viability in 38 months (Rodrigo, *1953*).

For safe sealed storage at room temperature, parsnip seeds must be dried to less than 1.7-percent moisture before being sealed (Joseph, *1929*). For safe storage at 5° to 7° C, the critical moisture level was 6.13 percent. When ovendry, 4-, 6-, and 8-percent moisture content parsnip seeds were stored for 6 years in sealed and unsealed containers at room temperature, 5°, and 6.7°, the best keeping was obtained with 4-percent moisture seeds at 6.7°, followed by ovendry at 5° (Beattie and Tatman, *1950*). Sealed storage also was effective in prolonging the life of carrot seeds (Coleman and Peel, *1952*; Dutt and Thakurta, *1956*). In open storage viability was lost in 9 months, but in sealed storage the seeds retained their viability.

Sealed storage prolonged the life of seeds of brussels sprouts, cabbage, cauliflower, knolkol or kohlrabi, radish, and rutabaga or swede (Coleman and Peel, *1952*; Rodrigo, *1953*; Dutt and Thakurta, *1956*; Madsen, *1957*). Loss of viability of dry seeds in sealed containers in a laboratory occurred in 37, 39, and 46 months for cabbage, radish, and cauliflower in that order (Rodrigo, *1953*). In open storage, seeds of cabbage, cauliflower, and kohlrabi lost viability in 9 months, but seeds sealed in a desiccator retained their viability for 8 years (Dutt and Thakurta, *1956*). Four lots of swede seeds retained their original germination for 10 years when stored at 4-percent moisture in sealed glass bottles (Madsen, *1957*). Cauliflower seeds stored similarly retained good viability for 8 years, whereas brussels sprouts seeds sealed at 4 percent retained their viability well for 16 years. Radish seeds with 4-percent moisture in sealed glass bottles retained their viability well for more than 8 years. Two lots declined slightly by the end of the 15th year.

For dry eggplant seeds in a sealed container in a laboratory, loss of viability occurred in 53 months (Rodrigo, *1953*). Tomato seeds with 5.9-percent moisture did not change significantly in viability and vigor during 36 months of sealed storage at 32° C. With increased seed moisture, temperature became more critical and tomato seeds lost viability rapidly above 21° (Asgrow Seed Co., *1954*). Potato seeds stored in sealed containers at temperatures from 0° to that of a basement room near Greeley, Colo., retained good germination for up to 13 years (Clark, *1940*; Wollenweber, *1942*; Stevenson and Edmundson, *1950*). Germination of seeds at room temperature dropped to 26 percent after 18 years and 17 percent after 20 years (Wollenweber, *1942*).

Use of sealed containers can extend the storage life of lettuce seeds (Coleman and Peel, *1952*; Bass et al., *1962*). Air-dry seeds in sealed containers in a laboratory lost viability in 27 months (Rodrigo, *1953*), but 4-percent moisture seeds retained good viability for 8 years at 21° C or colder (table 24). Seeds with 7-percent moisture lost all viability in less than 1 year at 32° and germinated only 4 percent after 4 years at 21° and 1 percent after 8 years at 10°. Seeds sealed with 10-percent moisture retained good germination for 8 years when stored at −12 and −1°. Seeds at 10° kept well for 3 years but deteriorated rapidly thereafter.

For dry asparagus seeds stored in sealed containers in a laboratory, loss of viability occurred at 50 months (Rodrigo, *1953*). Beet seeds placed in sealed bottles at 4-percent moisture and stored at room temperature did not lose viability in 10 years (Madsen, *1957*), and storage in tin cans was superior to storage in waxed cellophane, chlorinated rubber, or paper bags in linen ones (Coleman and Peel, *1952*). Pepper seeds canned with 4-percent moisture or less stored well for 36 months at 32° C or lower. As seed moisture content was increased, the maximum safe storage temperature decreased (Asgrow Seed Co., *1954*). Dry pepper seeds in sealed containers in a laboratory lost viability at 65 months (Rodrigo, *1953*). Okra seeds with 8- to 25-percent moisture content were stored for 10 years in sealed containers at room temperature of 21°–38° and at 2°–5° in cold storage. The upper moisture limit for 10 years in cold storage was 12 percent, and a moisture content as low as 8 percent was required for 7 years of safe storage at room temperature (Martin et al., *1960*).

Safe Moisture Levels for Sealed Storage

It is apparent from this discussion that simply sealing seeds in airproof and moistureproof containers is not necessarily a satisfactory packaging procedure for safe long-term storage. The research reviewed points up the need for careful drying to a moisture level that is safe for the highest temperature under which the seeds may be stored.

The results of Oregon experiments indicated that the following seed moisture level percentages can be considered safe for 3 years of sealed moistureproof storage under moderate temperatures for the following crops: Alfalfa 6, trefoil 7, clover, perennial ryegrass, soybean, and sweet corn 8, bentgrass, bluegrass, fescue, timothy, and vetch 9, and barley, bromegrass, common ryegrass, field corn, oats, rye, and wheat 10 (Ching, *1959*).

Moisture content percentages of the following are considered safe for up to 3 years of sealed storage: Cabbage, cauliflower, pepper, and tomato 5, celery and lettuce 5.5, cantaloup, cucumber, eggplant, onion, and watermelon 6, parsley 6.5, carrot and pea 7, beet 7.5, and bean, lawn grasses, spinach, and sweet corn 8 (Bass et al., *1961*).

Koopman *(1963)* listed the following percentages of moisture as safe for moistureproof packaging of flower seeds: *Ageratum* 6.7, *Alyssum* 6.3, *Antirrhinum* 5.9, *Aster* 6.5, *Bellis* 7.0, *Campanula* 6.3, *Lupinus* 8.0, *Myosotis* 7.1, *Nemesia* 5.7, *Penstemon* 6.5, *Petunia* 6.2, and *Phlox* 7.8.

The Federal Seed Act and all State seed laws require that all seed lots offered for sale be retested for germination at specified intervals. California took the lead in extending the germination test interval for adequately dried seeds marketed in hermetically sealed containers (Calif. Dept. Agr., *1973*). Federal regulations soon followed California's lead. The rules and regulations under the Federal Seed Act (U.S. Dept. Agr., *1968, pp. 17–83*) currently provide that the seeds in hermetically sealed containers shall not exceed the following percentages of moisture on a wet weight basis:

Agricultural Seeds

	Percent		*Percent*
Beet, field	7.5	Fescue, red	8.0
Beet, sugar	7.5	Ryegrass, annual	8.0
Bluegrass, Kentucky	6.0	Ryegrass, perennial	8.0
Clover, crimson	8.0	All others	6.0

Vegetable Seeds

	Percent		*Percent*
Bean, garden	7.0	Leek	6.5
Bean, lima	7.0	Lettuce	5.5
Beet	7.5	Muskmelon	6.0
Broccoli	5.0	Mustard, India	5.0
Brussels sprouts	5.0	Onion	6.5
Cabbage	5.0	Onion, Welsh	6.5
Cabbage, Chinese	5.0	Parsley	6.5
Carrot	7.0	Parsnip	6.0
Cauliflower	5.0	Pea	7.0
Celeriac	7.0	Pepper	4.5
Celery	7.0	Pumpkin	6.0
Chard, Swiss	7.5	Radish	5.0
Chives	6.5	Rutabaga	5.0
Collards	5.0	Spinach	8.0
Corn, sweet	8.0	Squash	6.0
Cucumber	6.0	Tomato	5.5
Eggplant	6.0	Turnip	5.0
Kale	5.0	Watermelon	6.5
Kohlrabi	5.0	All others	6.0

Moisture-Resistant Materials

Polyethylene Films.—The most extensively used thermoplastic film is made from an entire family of aliphatic hydrocarbon resins. Commercially available polyethylene resins fall into three groups: (1) Conventional low density types with specific gravity of 0.914 to 0.925 gm per cubic centimeter, (2) medium density types with specific gravity of 0.93 to 0.94, and (3) high density types with specific gravity of 0.95 to 0.96. Density differences are due to differences in molecular structure. Molecular structure determines the physical structure of the resins. Resin properties and extrusion variables determine film properties, which in turn determine the utility of the film. Physical properties determine the usefulness of a given film. They include tensile, tearing, and bursting strength, moisture vapor, carbon dioxide, and oxygen transmission rates, sealability, elongation, and folding endurance.

Conventional low density films have always been considered more satisfactory for seed packages than medium and high density films because of differences in bursting and tearing strength and film elongation or stretch. However, a special medium density film shows considerable promise. Resin manufacturers and film processors are continually working to improve bursting and tearing strength and elongation of medium and high density films.

Medium and high density films tend to show progressively less permeability to moisture vapor and gases than conventional low density films. Medium density films show one-half to one-third and high density films one-fourth to one-fifth the permeability of low density films. A 1-mil low density film tested at 37.8° C and 100-percent relative humidity will permit passage of 1.4 gm of moisture vapor through 100 square inches of film during 24 hours, a medium density film will transmit 0.7 gm, and a high density film 0.3 gm under the same conditions. A 10-mil low density film at 37.8° and 100-percent relative humidity will transmit 0.13 gm of moisture vapor per 100 square inches per 24 hours, approximately one-tenth the amount transmitted by the 1-mil film.

A medium density (specific gravity 0.938) polyethylene film has been developed that surpasses the performance of conventional polyethylene. A 7-mil film of this special material has a moisture vapor transmission rate of 0.10 gm per 24 hours per 100 square inches, which is less than that of 10-mil conventional polyethylene. This special medium density film has better tensile properties and greater elongation than conventional polyethylene. Because of its high percentage of stretch, this medium density film has very good puncture resistance.

Both clear conventional polyethylene and the special translucent, white, medium density polyethylene films are subject to slow deterioration on direct exposure to strong sunlight or ultraviolet radiation. However, deterioration can be retarded by incorporating into the film carbon black or other pigments, which will absorb the ultraviolet rays. The special medium density film has a very high resistance to stress cracking, which has been reported in a few instances with conventional polyethylene.

Rats and mice sometimes present a problem with conventional polyethylene, but no rodent attack on bags made of the special medium density material has been reported. Perhaps this material is a solution to rodent problems.

With a tight closure, such as is produced with a heat seal, both the 10-mil conventional polyethylene and 7-mil medium density polyethylene bags are almost completely insectproof. Thin films may be penetrated by some insects.

Polyethylene films can be laminated to themselves, to other films, paper, textile fabrics, and fiberboard. Moisture barrier and other physical properties are improved by laminations. The various properties of each film included in a laminate are more or less additive. Some laminated films are completely impervious to various gases and practically impervious to moisture vapor. Some laminated materials handle well on automatic packaging machinery and others handle best by hand depending on the nature of the materials used in the lamination.

Polyester Films.—These films are heat sealable, transparent, flexible plastic materials, with low moisture vapor, carbon dioxide, and oxygen transmission rates and great tensile strength. They will not dry out or become brittle with age because they contain no plasticizer. Polyester film can be laminated to itself and practically any other material, and its flexible laminates can be used with most flexible packaging equipment. A new construction that utilizes a base of light cotton fabric and metalized polyester film offers easier fabrication, stronger seals, resistance to flex damage, rough handling, and pinholes.

Polyvinyl Films.—These films are heat sealable, deteriorate slowly in sunlight, and have outstanding tensile strength and tear resistance, but they provide only moderate protection unless they are laminated to an effective moisture-barrier material. They heat-seal over a wide range of temperatures, are ideal for automatic packaging machinery, and laminate well to paper, foil, or other films. Polyvinyl's resistance to sunlight and aging indicates that packages will not dry out or become brittle.

Cellophane.—This family of films is made of regenerated cellulose and is produced in more than 100 varieties, each designed for a specific purpose. Moistureproof types, which have very low moisture vapor transmission rates, are used for small seed packages. Cellophane alone does not make a very satisfactory seed package because it becomes brittle with age or under arid conditions and it breaks easily. Several firms dealing in flexible packaging materials, however, do produce combinations of cellophane and polyethylene, which do not become brittle like cellophane alone and offer fairly good moisture protection. Polyethylene-cellophane laminates are rather extensively used in seed packaging as they heat-seal easily and perform well on automatic packaging machines.

Pliofilm.—Pliofilm is a thermoplastic, rubber, hydrochloride, plastic film, resistant to ripping, tearing, and splitting. It seals well at low temperatures, has good moisture-barrier properties, and can be laminated to itself, paper, foils, or other films. Pliofilm can be used on any package machine designed for flexible film packaging.

Aluminum Foil.—Annealed aluminum foil has a tensile strength of 8.5 pounds per inch of width per mil thickness. It increases in strength as the gage, or thickness, is increased and as the temperature is lowered. Tensile strength and resistance to tearing and bursting are greater for strain hardened foil than for annealed foil of the same thickness.

Aluminum foil has a low moisture vapor transmission rate, even for less than 1.5-mil thicknesses, which have tiny perforations called pinholes. These seem to be an inevitable result of rolling metal to very thin gages. Microscopic measurements of all the pinholes in 100 square inches of 0.4-mil foil gave an estimated area of 0.00004 square inch. A single hole of this area would transmit about 0.19 gm of water vapor per 24 hours at 37.8° C and 100-percent relative humidity. The number and size of pinholes decrease with increasing foil thickness. Moisture vapor transmission decreases also with increasing foil thickness. A 0.35-mil foil will transmit approximately 0.29 gm of water vapor per 100 square inches of foil per 24 hours at 37.8° and 100-percent relative humidity, a 0.5-mil foil will transmit 0.12 gm of water vapor under the same conditions, and thicker films transmit almost no water vapor at all.

Aluminum foil is commonly used in laminations and separately as a coating and overwrap material for cardboard boxes. Aluminum foil alone does not make good seed packages, but it can be bonded to other materials to produce combinations having almost any desired characteristics. Even though thin gages of aluminum foil have some pinholes, combinations with various supporting materials, such as paper or plastic films, offer effective barriers to moisture vapor and gas transfer. With the proper selection of materials, combinations can be produced that will completely restrict moisture vapor transfer.

Laminations.—Laminations of aluminum foil with other materials have been satisfactorily used for all sizes of seed packages. Examples of such laminations include (1) aluminum foil/glassine paper/aluminum foil/heat-sealing lacquer, (2) aluminum foil/tissue paper/polyethylene, and (3) kraft paper/polyethylene/aluminum foil/polyethylene.

Burlap and cotton can be laminated to paper with asphalt or a compound latex adhesive to produce a material that will provide good protection from liquid water but only limited protection from water vapor. Other constructions utilize such barrier materials as vegetable parchment, pliofilm, polyethylene, rubber coatings, and foil. Moisture-resistant multiwall bags usually have a barrier material, such as asphalt or polyethylene, between the two outer layers of paper. A good asphalt laminate at 26.7° C and 75-percent relative humidity has a moisture vapor transmission rate of 0.17 gm per 100 square inches per 24 hours. Some seed bags are constructed of paper/polyethylene/aluminum foil laminates. These combinations afford better moisture protection than either foil or polyethylene used along with paper. Films of cellophane, pliofilm, polyester, polyvinyl, aluminum foil, and polyethylene are used alone or in various combinations for seed packages.

Testing Moisture Vapor Transmission of Flexible Materials

There are several methods of measuring the rate of moisture vapor transmission through a material. Two methods used in Norway are the dish method and the method according to Bange (Fornerod, *1963*). In the dish method a strong desiccant is sealed in a special dish by a piece of the material to be tested. The dish is placed at a controlled temperature and relative humidity. At periodic intervals the dish is weighed and the weight change is used to calculate the rate of passage of moisture vapor through the test material.

The method according to Bange is similar but more accurate for very slow moisture vapor transmission. The Bange method uses a salt solution rather than a desiccant in the dish, which is placed in a measuring apparatus. A strong desiccant is placed in the left pan of the balance and its weight gain is measured from time to time. From these measurements the moisture vapor transmission is calculated.

An easy method for an individual to use is to fill a container of the material to be tested with water, seal it, and place it in a desiccator over calcium chloride. Moisture penetration is measured by the weight loss of the package. Conversely, the package can be sealed full of calcium chloride, then placed in a desiccator over water. In this case weight gain is used to measure the rate of moisture vapor penetration. No elaborate special equipment is required for this method of testing moisture vapor transmission, only a reliable balance for weighing.

From the test data available, not all manufacturers appear to use the same temperature and relative humidity when testing moisture vapor transmission of their materials. Some give grams of water per 100 square inches of test material per 24 hours at 37.8° C, with 100-percent relative humidity on one side and 0 percent on the other side of the material being tested. Others use the U.S. Bureau of Standards method, which measures water vapor penetration as grams of water per 100 square inches of test material per 24 hours at 37.8°, with 90-percent relative humidity on one side and 0 percent on the other. For law enforcement purposes the Federal Seed Act specifies use of the Bureau of Standards method.

In the National Seed Storage Laboratory at Fort Collins the moisture protection of test materials is measured by the change in seed moisture content over time. Seeds in packages made of good moisture-barrier materials show little change in seed moisture, whereas seeds in poor moisture-barrier materials gain or lose moisture rapidly. A simple method, which does not destroy the seeds, as when making a seed moisture test, is to weigh the test package when filled and at intervals thereafter. However, the outside of the package must be free of liquid water. Also, paper on the outside can hold extra water and give erroneous results. The moisture vapor transmission rate can be calculated using the amount of weight change, the time between weighings, and the surface area of the package. The moisture vapor transmission rate measured by either of the last two methods will be less than the rate for the same material measured by the U.S. Bureau of Standards method unless the test package is subjected to the specified temperature and relative humidity.

Moisture-Barrier Storage

The development of polyethylene and other flexible moisture-barrier packaging materials prompted studies to determine the value of these materials as protective packages for seeds. Thin gages of polyethylene, polyester, and similar materials do not provide very much moisture protection (Barton, *1949, 1953*; Isely and Bass, *1960*; Miyagi, *1966*). Storage life was increased by storage in containers made of 5- or 10-mil polyethylene for the following seeds: Alta tall fescue, Chewings fescue, corn, creeping red fescue, cucumber, hemp, Highland bentgrass, kenaf, Kentucky bluegrass, onion, peanut, ryegrass, sudangrass, and wheat (Anonymous, *1959*; Cooper, *1959*; Bass, *1960, 1968*; Ching et al., *1960*; Toole et al., *1960, 1961*; Milligan and Hayes, *1962*; and Grabe and Isely, *1969*).

Kraft multiwall bags with laminated paper/foil/polyethylene liners were as effective as tin cans in maintaining a moisture barrier during storage and shipment of cabbage, soybean, and wheat seeds (Caldwell, *1962*). The water vapor transmission rates of polyethylene materials are inversely proportional to their protective value for seed storage under tropical, temperate, and warehouse conditions (Ching and Abu-Shakra, *1965*). Packaging materials containing aluminum foil provide good moisture protection (Harrington, *1960a, 1960b, 1963*; Lowig, *1963*; Bass and Clark, *1974*; and Clark and Bass, *1975*).

The need for moisture protection is determined by the storage conditions and so is the protective value of a barrier material to a certain extent. A good barrier material, such as a heavy foil laminate, will provide adequate moisture protection regardless of the storage conditions; a mediocre material will not give satisfactory moisture protection when relative humidity is high (table 25).

To provide a rigorous test, we stored crimson clover seeds in a variety of materials (table 26) in a 20° C night (15 hours) – 30° day (9 hours), water-curtain germination chamber, in which the relative humidity was 95 to 100 percent constantly. The data (table 25) show which materials provided satisfactory moisture and germination protection. Seeds in the same materials stored at 10° and 50-percent relative humidity showed a different moisture and germination picture.

TABLE 25.—*Moisture content and germination of crimson clover seeds packaged in various flexible materials and stored for 2 years in walk-in germinator held at 20°–30° C and 95- to 100-percent relative humidity (RH) or at 10° and 50-percent relative humidity*

Material No.[1]	Moisture when stored	Moisture after 2 years at—		Germination when stored	Germination after 2 years at—	
		20°–30° and 95- to 100-percent RH	10° and 50-percent RH		20°–30° and 95- to 100-percent RH	10° and 50-percent RH
	Percent	*Percent*	*Percent*	*Percent*	*Percent*	*Percent*
1	3.4	[2]27.9	8.9	74	[2]0	72
2	3.4	13.0	4.6	74	42	73
3	3.3	20.0	6.4	74	0	76
4	3.3	18.7	6.7	74	0	85
5	3.4	29.0	6.9	74	0	79
6	3.4	22.5	8.5	74	0	77
7	3.4	18.5	5.8	74	1	70
8	3.3	16.2	5.8	74	0	84
9	3.3	10.3	5.8	74	72	80
10	3.4	3.4	3.3	74	75	81
11	3.3	6.6	3.3	74	72	74
12	3.3	19.8	7.0	74	0	79
13	3.4	13.1	5.7	74	27	79
14	3.2	3.3	3.4	74	73	70

See footnotes at end of table.

TABLE 25.—*Moisture content and germination of crimson clover seeds packaged in various flexible materials and stored for 2 years in walk-in germinator held at 20°–30° C and 95- to 100-percent relative humidity (RH) or at 10° and 50-percent relative humidity*—Con.

Material No.[1]	Moisture when stored	Moisture after 2 years at—		Germination when stored	Germination after 2 years at—	
		20°–30° and 95- to 100-percent RH	10° and 50-percent RH		20°–30° and 95- to 100-percent RH	10° and 50-percent RH
	Percent	*Percent*	*Percent*	*Percent*	*Percent*	*Percent*
15	3.4	3.5	3.5	74	78	77
16	3.3	3.4	3.3	74	73	78
17	3.3	[2]30.9	8.8	74	[2]0	80
18	3.4	4.0	3.3	74	70	81

[1]See table 26 for identification of materials.
[2]Seed dead by 6-mo test.

Results of our studies suggest that a foil laminate such as material 10, 11, 14, 15, 16, or 18 (tables 25 and 26), properly sealed, provides as much moisture protection as a sealed metal can. The data (table 25) also show that for long-term storage the container must provide adequate moisture protection for the most humid condition to which the seeds could be subjected, not just the intended storage relative humidity. When preserving germ plasm, care must be taken to guard against any eventuality.

TABLE 26.—*Construction and source of 18 flexible packaging materials tested for suitability as moisture-barrier seed packages*[1]

Material No.	Construction	Source
1	20-lb brown kraft coin envelopes	Cupples-Hess Co., Div. of St. Regis Paper Co.
2	30# paper/17# Saran/0.5-mil polyethylene	Chase Bag Co.
3	0.5-mil mylar/Saran coating/2-mil polyethylene	Continental Can Co.
4	0.5-mil bioriented polypropolene/adhesive/ 250K204F cellophane/28.8-lb (2-mil) polyethylene extrusion coating.	Milprint, Inc.
5	250K204F cellophane/adhesive/30- to 39-lb Saran-coated opaque glassine.	Do.
6	0.5-mil A mylar/29.2-lb medium density polyethylene extrusion coating.	Do.
7	1.25-mil oriented polypropylene, 1-side coated with 2-mil branched polyethylene.	North American Packaging Co.
8	4-mil cast polypropylene	Avisun Corp.
9	Heat-seal coated 2-mil foil	Aluminum Co. of America.
10	Paper/foil/polyethylene designated FX12530	Reynolds Metals Co., Packaging Research Div.
11	Polypropylene/foil/polyethylene designated FX12529.	Do.
12	0.5-mil oriented polypropylene/Saran coating/2-mil polyethylene.	Continental Can Co.
13	Marlex[2] Tr 101 6-mil high density polyethylene	Mehl Manufacturing Co.

See footnotes at end of table.

TABLE 26.—*Construction and sources of 18 flexible packaging materials tested for suitablility as moisture-barrier seed packages*[1]—Con.

Material No.	Construction	Source
14	45-lb polyethylene/0.5-mil foil/10-lb polyethylene/50-lb extensible multiwall kraft.	Union Bag Camp Co.
15	2.5-oz sq yd scrim/1-mil polyethylene/0.35-mil foil/3.0-mil polyethylene.	Chase Bag Co.
16	0.5-mil A mylar/7-lb polyethylene/0.35-mil foil/15-lb polyethylene.	Rap Industries.
17	5-mil ethyl-vinyl-acetate film	Union Carbide Corp., Plastics Div.
18	140 RSO cellophane/10-lb polyethylene/0.35-mil foil/17-lb polyethylene.	Rap Industries.

[1]These packaging materials were contributed through the courtesy of the individual firms.

[2]A trademark name for Phillips family of olefin polymers.

Use of Desiccants in Sealed Containers

Although the use of desiccants has received some attention from seed researchers, it has not received much attention from the seed industry. The longevity of rice seeds in sealed containers has been increased by including in the container a drying agent, such as calcium oxide (quicklime) or calcium chloride (Nakajima, *1927*; Kondo and Kasahara, *1941*; Ito and Hayashi, *1960*). Rice seeds in a closed container with calcium chloride retained good viability for 9 and 10 years (Nakajima, *1927*; Kondo and Kasahara, *1940*). Germination of rice seeds having 10-percent moisture when sealed was maintained for 1 year at 40° C when calcium chloride was enclosed with the seeds. Seed with 16-percent initial moisture was preserved at room temperature with calcium chloride, and without it all viability was lost (Kondo and Okamura, *1931*).

Tobacco seeds sealed with calcium chloride germinated 88 percent after 11 years and 81 percent after 25 years (Kincaid, *1943, 1958*), and seeds sealed with calcium oxide germinated 87 to 92 percent after nearly 21 years (Schloesing and Leroux, *1943*). Storage with silica gel improved longevity of timothy seeds (Roberts, *1962*), quicklime preserved *Viola tricolor* seeds (Lowig, *1953*), but seeds of *Zoysia japonica* sealed with calcium chloride lost germination rapidly after 2½ years along with seeds sealed without a desiccant (Radko, *1955*). Temperature had little effect on the longevity of bean, eggplant, lettuce, and tomato seeds stored over calcium chloride (San Pedro, *1936*). Germination percentages of several crops after 10 years over calcium chloride were *Avena sativa* 89, *Capsicum annuum* 70, *Cucumis sativus* 80, *Cucurbita pepo* 55, *Oryza sativa* 62, *Raphanus sativus* 81-89, *Solanum melongena* 37, and *Zea mays* 79 (Kondo and Kasahara, *1940*).

Most desiccants do not injure seeds; however, quicklime did have an injurious effect on seeds of several species, such as *Oryza sativa, Phaseolus radiata* var. *pendulus* (= *Vigna radiata*), and *Vicia sativa* (Nakajima, *1927*). Calcium oxide did not improve the longevity of 'Golden Cross Bantam' sweet corn (Edmond, *1959*).

Although most of the studies showed that seeds sealed with desiccants lived somewhat longer than those sealed without them, their use has not received wide application in the seed industry because of several factors. Probably the most important are the added cost of the desiccant and the larger container required for it. Also, when a package is opened, the desiccant is soon useless. Because most seeds are stored commercially only from harvest to the next planting season, the added expense is not easily justified except possibly in very humid areas where seeds deteriorate rapidly under normal storage conditions.

Use of desiccants may offer some economic advantage for long-term storage. However, without the results of a comprehensive study, one cannot be certain of their value in preserving seed viability and vigor during long-term storage in sealed moistureproof containers, because for adequately dried seeds a desiccant may be of no benefit. Such a study should include several kinds of seeds with a wide range of moisture levels, stored with different amounts of each kind of desiccant, and at temperatures from below freezing to 32° C or higher. Desiccants could be very beneficial when used in conjunction with some of the flexible packaging materials that have limited moisture-barrier qualities.

MONITORING SEED STORAGE ENVIRONMENT AND SEED CONDITION

Germination and Viability

Ordinarily seeds are tested for germination and viability by laboratories equipped and staffed to determine the quality of seeds. Inexperienced individuals planning to conduct germination and viability research should receive some training in a reputable laboratory. In the germination test a seed to be considered as having germinated must produce a seedling with normal or approximately normal features. Some kinds of plants produce hard seeds that are considered to be viable although they do not germinate when tested according to officially accepted procedures. Sometimes dormant seeds require special germination test procedures. A viability test aims to quickly detect all live seeds whether dormant or not. The tetrazolium and the embryo excision tests are used for this purpose.

The procedures for testing seed germination and viability are found in the "Rules for Testing Seeds" by the Association of Official Seed Analysts *(1970)*, special handbooks published by the Association of Official Seed Analysts, the "International Rules for Seed Testing" published by the International Seed Testing Association *(1966)*, the "Rules and Regulations of the Secretary of Agriculture," U.S. Department of Agriculture, published in the Federal Seed Act of August 9, 1939, with revisions, and in certain supplementary publications (Grabe, *1970*; Justice, *1972*). Information concerning the availability of these rules for seed testing and the special handbooks can be obtained from the Seed Branch, Grain Division, Agricultural Marketing Service, U.S. Department of Agriculture, Washington, D.C. 20250.

Seed Vigor

Vigor as applied to seeds is an indefinite term. No definition for seed vigor has been accepted generally. The methods used to test seed vigor are just as vague. Neither the Association of Official Seed Analysts nor the International Seed Testing Association has included methods of testing seeds for vigor in their procedures. The best known and commonly used vigor test is the so-called cold test developed for corn and subsequently adapted for a few other crop species. One problem with this test has been the difficulty of standardizing the fungi and soil used in making the test. Additional information on this test can be obtained from the Iowa State University Seed Laboratory, Ames 50010, where the test was developed.

Other procedures used in research include the respiration test, which measures oxygen consumption and carbon dioxide release, the GADA (glutamic acid decarboxylase activity) test, various types of stress tests, rate of seedling growth test, and the tetrazolium test. Each of these tests has been useful for specific kinds of seeds but has not proved reliable for a wide range of various kinds of seeds. Woodstock *(1973)* has reviewed the literature on vigor tests.

Seed Moisture Content

Practical methods of testing seeds for moisture include oven methods and electric moisture meters. Basically the oven methods operate on the principle that the seed moisture is driven off by heating. The difference in weight of the seeds caused by heating represents the moisture content. The temperature and time of heating have been reported for many kinds of seeds. The heating regimes have been developed to minimize weight changes not produced by moisture.

Large seeds are frequently ground to decrease drying time. Many seeds are heated to 130° C, whereas others should not be heated above 100° and a few at lower temperatures. A common method is to dry seeds for 24 hours at 100° to 105°. At the higher temperatures, volatile materials can be driven off and oils and fats oxidized, both resulting in weight changes. As the drying temperature is decreased, the drying time must be increased proportionately. Drying schedules have been published in the "International Rules for Seed Testing" *(1966)*. These rules were developed for air ovens that operate at normal atmospheric pressure. Vacuum ovens with the advantage of reducing drying time are available. All oven methods have the disadvantage of requiring much equipment, weighing of materials, and considerable time for testing. However, they are the more accurate of the practical methods.

Electric moisture meters have an advantage in speed over all other practical methods. Most electric moisture meters are based on measurements of either conductivity or the dielectric properties of the seed. Although these measurements vary with the moisture content of the seeds, they are affected to some extent by other factors. Calibration charts have been developed for many kinds of seeds. Calibrations should be based on the testing of relatively large numbers of samples, covering a wide range of moisture contents, sources of samples, and years of harvest.

With most moisture meters, the test can be completed within 1 or 2 minutes. Disadvantages are relatively high cost of equipment, need for careful, painstaking calibration for each kind of seed, and in some cases, especially for very wet and very dry seeds, failure of the equipment to be highly accurate. For more information on methods of testing seeds for moisture content, see Hlynka and Robinson (*in* Anderson and Alcock, *1954*), Hart et al. *(1959)*, and Zeleny *(1961)*.

Fungi and Bacteria

Methods of testing seeds for the presence of seedborne micro-organisms are given in the "International Rules for Seed Testing" *(1966)*. For additional information, see Muskett *(1948)*, de Tempe *(1961)*, Naumova *(1970)*, Baker *(1972)*, and Neergaard *(1973)*.

Relative Humidity

A psychrometer is commonly used to measure relative humidity. Sling psychrometers are inexpensive, reasonably accurate, and easy to use. This instrument consists of a dry-bulb thermometer, which measures the ambient temperature, and a wet-bulb thermometer, which measures the reduced temperature caused by the cooling effect of the evaporating liquid. This cooling effect is in proportion to the relative saturation of the air being tested. Recording psychrometers have the advantage of providing a record at any time over the interval covered by the clock and chart. Both temperature and relative humidity are recorded by the hygrothermograph. This instrument requires special attention when changing the charts. The directions for its use should be followed very closely. Recording units are available that use electrical resistors to measure humidity and thermistors to measure temperature. Although expensive, they are very accurate. Major changes in relative humidity can be detected with cobalt chloride cards. At different positions on the card, the color changes as the relative humidity progressively increases.

Temperature

Thermometers, like psychrometers, are made in several forms. When a high degree of accuracy is not required, simple, inexpensive stem-type thermometers can be used to measure the temperature of the air or the mass of seeds. Although some or most of the thermometers of this type may be accurate, they must be removed from the seed mass before taking the reading. By the time the reading is taken, the mercury column may have changed, and thus it gives an erroneous reading. Recording thermometers, with or without a sensing element, are available. Of course, these instruments provide a continuing record, which can be cited for future reference if desired. Multipoint recorders provide essentially the same information, but each instrument can be equipped with several sensing units, or thermocouples, allowing the monitoring of several locations simultaneously. Thus, hot spots in a mass of seed or grain may be detected with a thermocouple that conceivably could go undetected if the other temperature sensing and recording instruments were used. Thermographs, recording thermometers, and multipoint recording units are rather expensive. Their use may or may not be justified depending on the degree of accuracy desired, chance of product heating in storage, extent of business operation, and frequency of use.

Storability of Seed Lots

Considerable research, mostly at the Mississippi State University Seed Technology Laboratory, has been conducted to develop a practical method for predicting relative storability of seed lots, especially those to be stored for future use. Although several potential tests were devised and evaluated in this research, two were more successful than others—the accelerated aging test and the modified aging test.

The following information has been taken from reports by Delouche et al. (*1967, 1968*), Delouche and Baskin *(1971)*, and Goff *(1972)*. "In the accelerated aging test, the germinative responses of representative samples of seed lots are determined after exposure to conditions of temperatures of 40° to 45° C and 100-percent relative humidity for varying periods of time up to 204 hours. The modified aging test measures the germinative responses of samples after they have been held at 30° and 75-percent relative humidity for various periods of time."

Optimum accelerated aging conditions and treatment times in the 2 tests are given for 16 kinds of seeds in table 27.

TABLE 27.—*Optimum accelerated aging conditions and treatment times in 2 tests for predicting seed storability of 16 crops*[1]

Crop	Accelerated aging test		Modified aging test
	Temperature at 100-percent relative humidity	Treatment time	Treatment time at 30° C and 75-percent relative humidity
	°C	*Hours*	*Weeks*
Alfalfa	42	84	6
Bean	42	84	20
Bromegrass	45	72	9
Corn	42	84	24
	45	60	---
Crimson clover	40	72	9
Fescue	42	84	15
Lespedeza	40	72	6
Lettuce	40	72	9
Onion	42	72	6
Radish	45	48	24
Red clover	40	72	9
Sorghum	45	72	12
Soybean	40	72	9
Timothy	42	60	6
Watermelon	45	144	21
Wheat	45	48	18

[1] Data from Delouche and Baskin *(1971)*.

We do not claim that these tests will accurately predict the storability of all seed lots of a given crop species. The principles involved do offer the seedsman a means of detecting some seed lots that should not be stored for extended periods of time. More accurate predictions could be made if the seed lots were stored under controlled temperature and relative humidity rather than under ambient conditions.

The researchers who developed the accelerated aging test of predicting seed storability are confident that the method or modifications will ultimately be perfected so that storability of seed lots can be predicted with considerable accuracy. Additional information on these test procedures can be obtained from the Seed Technology Laboratory, Mississippi State University, State College 39762.

Detection and Identification of Fungi and Insects

Ordinarily the manager or superintendent of the storage warehouse must monitor his seed stocks. It is his responsibility to check for evidence of fungi and insects. Their presence may be detected either by observing their growth on or in the seed mass or by development of hot spots. If hot spots are detected, the cause should be ascertained immediately. The cause may be traced to molds, insects, or both. Specialists may be required to identify molds or insects. The State agricultural colleges and universities and State departments of agriculture frequently will make simple identifications of occasional specimens for residents. Depending on their background, some county agricultural agents may perform this service. Seed companies often employ a plant pathologist. Some official and commercial seed testing laboratories routinely conduct seed health tests.

SOME PRACTICAL INFORMATION FOR STORING AND TRANSPORTING SEEDS AT AMBIENT CONDITIONS

Examples of Storability of Different Plant Species

There is much variability in the lifespan or storage life of seeds of different plant species. The lifespan of seeds of certain aquatic and woody plants is very short, whereas seeds of some leguminous species and a few other kinds retain viability for over a hundred years. Only the practical storage life of cultivated species of field, vegetable, herb, and flower seeds is discussed here, not species with extremely short or extremely long storage lives. Seed longevity per se is not included. The information here relates to the retention of viability of most individual seeds in a lot, not the elapsed time before the last viable seed in a lot or sample is dead.

Species for which information is available have been listed. The data can be used as guidelines for closely related species not given. The information is rather scanty, some of it is outdated, and most of it has been obtained under nonuniform storage and test conditions. This obviates accurate comparisons.

As indicated by the column heading, "Relative Storability Index," in the tabulation, the data do not refer to specific years of storage. In general, the authors classified in category 1 those plant species of which 50 percent or more of the seeds can be expected to germinate after 1 to 2 years of storage under favorable ambient conditions at latitudes of approximately 35° to 48° N., in category 2 after 3 to 5 years, and in category 3 after 5 or more years.

Examples of relative storability of field, vegetable, herb, and flower seeds of high viability and vigor are as follows:

Seed — FIELD CROP SEEDS	*Relative storability index*
Alfalfa	3
Alyceclover	3
Bahiagrass	2
Barley	2
Bean, field	2
Beet, field and sugar	3
Bentgrass:	
Colonial (including Astoria and Highland)	3
Creeping	3
Velvet	3
Bermudagrass	1
Bluegrass:	
Bulbous	2
Canada	2
Kentucky	2
Nevada	2
Rough	2
Texas	2
Wood	2
Bluestem:	
Big	2
Little	2
Sand	2
Brome:	
Mountain	2
Smooth	2
Buckwheat	2
Buffalograss	3
Burclover:	
California	3
Spotted	3
Buttonclover	3
Canarygrass	1
Canarygrass, reed	1
Carpetgrass	3
Clover:	
Alsike	3
Berseem	2
Crimson	2

FIELD CROP SEEDS—con.

Seed	*Relative storability index*
Clover—Con.	
Ladino	3
Lappa	3
Large hop	3
Persian	3
Red	3
Strawberry	3
Subterraneum	2
Suckling (small hop)	3
White	3
Corn:	
Field	1
Pop	2
Cotton	1
Cowpea	1
Crested dogtail	2
Crotalaria	3
Dallisgrass	1
Dropseed, sand	2
Fescue:	
Chewings	2
Hair	2
Meadow	2
Red	2
Sheep	2
Tall	2
Flax	2
Grama:	
Blue	2
Side-oats	2
Hardinggrass	1
Hemp	2
Indiangrass, yellow	1
Japanese lawngrass	2
Johnsongrass	2
Kudzu	2
Lespedeza:	
Korean	1
Sericea or Chinese	2
Siberian	2
Striate	1
Lovegrass, weeping	2
Lupine:	
Blue	1
White	1
Yellow	1
Manilagrass	2
Meadow foxtail	2
Medic, black	3
Millet:	
Foxtail (common, German, golden, Hungarian, Siberian)	1

Seed FIELD CROP SEEDS—con.	*Relative storability index*
Millet—Con.	
Japanese	2
Pearl	1
Proso	1
Oatgrass, tall	1
Oats	2
Orchardgrass	1
Panicgrass, blue	2
Pea, field	2
Peanut	1
Poppy	2
Rape	2
Redtop	1
Rescuegrass	2
Rhodesgrass	2
Rice	2
Ricegrass, Indian	2
Roughpea	2
Rye	1
Ryegrass:	
Italian	2
Perennial	2
Sainfoin	1
Smilo	3
Sorghum: Grain and sweet	1
Soybean	1
Sudangrass	1
Sunflower	1
Sweetclover:	
White	3
Yellow	3
Switchgrass	2
Timothy	2
Tobacco	1
Trefoil:	
Big	2
Birdsfoot	2
Vaseygrass	1
Velvetbean	1
Vetch:	
Common	2
Hairy	3
Hungarian	2
Monantha	2
Narrowleaf	3
Purple	2
Woollypod	3
Wheat, common	2
Wheatgrass:	
Fairway crested	2
Intermediate	2

Seed	*Relative storability index*
FIELD CROP SEEDS—con.	
Wheatgrass—Con.	
Pubescent	2
Slender	2
Standard crested	2
Tall	2
Western	2
Wild rye:	
Canada	2
Russian	2
Zoysia (see Japanese lawngrass and manilagrass)	
VEGETABLE SEEDS	
Artichoke	1
Asparagus	1
Bean:	
Garden	1
Lima	1
Beet	3
Broadbean or horsebean	2
Broccoli	2
Brussels sprouts	2
Cabbage	2
Cardoon	1
Carrot	2
Cauliflower	2
Celeriac	2
Celery	2
Chicory	2
Collards	2
Corn, sweet	2
Cowpea	1
Cress:	
Garden	2
Water	2
Cucumber	2
Dandelion	1
Eggplant	2
Endive	2
Kale	2
Kohlrabi	2
Leek	1
Lentil	1
Lettuce	1
Muskmelon (cantaloup)	2
Mustard, India	2
Okra	2
Onion	1
Pakchoi	1
Parsley	1
Parsnip	1
Pea, garden	2
Pepper	1

Seed	Relative storability index
VEGETABLE SEEDS—con.	
Pe-tsai (Chinese cabbage)	2
Pumpkin	2
Radish	2
Rhubarb	1
Rutabaga	2
Salsify	1
Soybean	1
Spinach:	
Common	2
New Zealand	2
Squash	2
Tomato	3
Turnip	2
Watermelon	2
HERB SEEDS	
Anise	1
Balm	1
Basil, sweet	2
Borage	1
Caraway	1
Chervil	1
Coriander	1
Dill	1
Fennel	1
Hyssop	1
Marjoram	1
Rosemary	1
Sage	1
Savory	1
Thyme	1
FLOWER SEEDS	
Achillea, the pearl	2
Ageratum	2
Alyssum	2
Amaranthus	3
Anemone	1
Angel-trumpet	2
Arabis	1
Armeria	1
Asparagus, fern	1
Aster, China	1
Babysbreath	2
Bachelor's button, cornflower	2
Balloonflower	1
Balsam	2
Begonia	1
Bellflower, peach	2
Browallia	1

Seed	FLOWER SEEDS—con.	*Relative storability index*
Bugloss		2
Butterflyflower		2
Calceolaria		1
Calendula		2
Calliopsis, dwarf and tall		2
Candytuft:		
Annual		1
Perennial		1
Canna		2
Canterbury-bells		2
Carnation		2
Cathedral bells		1
Centaurea:		
Royal		3
Velvet		3
Chrysanthemum, annual		3
Cineraria, common		2
Cleome, spiderflower		1
Cockscomb		2
Cockvine		1
Coleus, common		2
Columbine		1
Coneflower		1
Coralbells		1
Coreopsis, perennial		1
Cosmos		2
Cyclamen		2
Cypressvine		3
Dahlia		2
Daisy:		
African		1
African lilac		2
English		1
Painted		1
Shasta		3
Swan river		2
Dames rocket, sweet rocket		2
Dusty-miller		2
Firebush, Mexican		1
Flax:		
Flowering		3
Perennial		3
Foxglove		1
Gaillardia		2
Geum		1
Gilia		2
Globe amaranth		2
Gloxinia, common		1
Godetia		2
Gourds		3

Seed	FLOWER SEEDS—con.	*Relative storability index*
Heliopsis		1
Heliotrope		1
Hibiscus		2
Hollyhock		3
Hyacinth-bean		2
Iris, Japanese		2
Jerusalem or Maltese cross		2
Jobs-tears		2
Lantana		1
Larkspur:		
Annual		1
Hybrids		2
Linaria		2
Lobelia		2
Lunaria, honesty		2
Lupine:		
Annual types		2
Russell hybrids		2
Marigold:		
African		2
French		2
Marvel of Peru, four-o'clock		2
Matricaria		1
Mignonette		1
Morningglory		3
Myosotis		1
Nasturtium		2
Nemesia		1
Nicotiana		2
Nigella		2
Pansy		1
Penstemon		1
Petunia		2
Phacelia		2
Phlox		1
Physalis		2
Pinks, China		2
Poppy:		
California		2
Corn, shirley		3
Iceland		2
Mexican tulip		1
Oriental		2
Portulaca		2
Primrose		1
Rose campion		2
Sage:		
Mealycup		2
Scarlet		1
Salpiglossis		3

Seed	FLOWER SEEDS—con.	*Relative storability index*
Saponaria		2
Scabiosa		2
Snapdragon		2
Sneezeweed, helenium		2
Snow-on-the-mountain		2
Solanum		2
Statice		1
Stocks		3
Strawflower		1
Sunflower		2
Sweetpea:		
Annual		3
Perennial		2
Sweet-william		2
Verbena		2
Vinca, periwinkle		1
Viola		1
Wallflower		2
Zinnia		3

The following references were used in compiling the list: Bodger Seeds Ltd. *(1932)*, Filter *(1932–33)*, Goss *(1937)*, Harrington *(1972)*, James et al. *(1964)*, Madsen *(1962)*, Owen *(1956)*, Pritchard *(1933)*, Sifton *(1920)*, Ullmann *(1949)*, Welton *(1921)*, and Wheeler and Hill *(1957)*.

Climatological Data Pertinent to Seed Storage

Although the relative humidity and temperature of the storage environment can be controlled artificially, the cost of providing such control in large storage areas is usually not economically feasible. Thus, most seeds stored in North America are kept at ambient temperatures and relative humidities. Storage as used here includes holding of seeds for processing, shipment, or sale, as well as the period of shipping.

Because of the wide range of climatic conditions in the United States, its territories, Puerto Rico, Canada, and Mexico, climatological data useful to seedsmen, seed merchants, and farmers could be provided in maps; however, weather maps give less specific information than tabular data and may be difficult to interpret.

The data in table 28 are relatively longtime averages and should provide reliable guidelines for persons planning to store seeds at ambient conditions at the indicated locations. Data are given for January, April, July, and October, the first month of each season, as well as annual averages. Locations have been selected to give a fairly representative cross section of the area covered. For locations not shown, data from the nearest station should be used or adapted, based on known facts about the local climate. For example, if one is going to store seeds at ambient conditions in Dayton, Ohio, from July to January, he would probably check the conditions at Cincinnati and Columbus, Ohio, as well as at Indianapolis, Ind. Six-month averages for these three locations for July and October are temperature 19° C and relative humidity 70.1 percent. For practical purposes, these values would be close enough for Dayton. If the seed is to be stored for a year, the annual averages for the three locations should be used, as 12° and 71.7-percent relative humidity. Other adaptations may be necessary for other locations, especially where relatively great climatological differences occur within short distances, as in mountainous areas or near large bodies of water. Annual averages should be used with great care because of extreme seasonal differences that can be masked in them.

These data provide a guide for drying carryover seed and storing it in moisture-barrier containers. It should be remembered that basement rooms, unless well ventilated, always have a higher relative humidity than aboveground rooms and can be potentially dangerous as seed storage areas.

Another possibility is to obtain original and up-to-date information from the National Weather Records Center, U.S. Department of Commerce, NOAA, Asheville, N.C. 28801. This Center can provide current, more complete, and longtime averages of temperatures, relative humidities, and several other weather and climatological data for many stations not shown in table 28.

Relation of Storage Conditions to Intended Storage Periods

The first question asked of seed storage specialists is "How should I store my seed?" For an adequate answer the specialist must know where and what kinds are to be stored for how long. In arid areas many kinds of seeds can be stored for 5 years or longer under ambient conditions provided the temperature is not too high. In temperate regions the same kinds of seeds may keep only 2 or 3 years, whereas in the humid subtropical and tropical regions viability may be lost in a few months or even weeks at ambient conditions.

A useful guide for commercial seed storage for up to 5 years is Harrington's *(1960d)* thumb rule that states "the sum of the temperature in °F and the percent relative humidity should not exceed 100." For example, various kinds of seed stored at 50° F (10° C) and 50-percent relative humidity showed no loss in viability during 5 years of storage (James et al., *1967*). Comparable results can be obtained by drying seed to 5-percent or lower moisture content and storing it in sealed moisture-barrier containers at ambient temperatures up to 32° C. At higher ambient temperatures some refrigeration is desirable.

TABLE 28.—*Climatological data for 183 weather stations in the United States, its territories, Puerto Rico, Canada, and Mexico*[1]

Location of weather station	Average temperature and relative humidity (RH)[2] for—									
	January		April		July		October		Annual	
	Temp.	RH	Temp.	RH	Temp.	RH	Temp.	RH	Temp.	RH
	°C	*Percent*	*°C*	*Percent*	*°C*	*Percent*	*°C*	*Percent*	*°C*	*Percent*
United States:										
Alabama:										
Birmingham	8	72	17	65	27	74	18	70	17	70
Mobile	11	74	19	73	28	78	21	72	20	73
Montgomery	9	74	18	67	28	75	19	70	19	74
Alaska:										
Anchorage	−11	74	2	66	14	73	2	77	2	72
Fairbanks	−24	69	2	61	16	70	−3	79	3	69
Juneau	−4	70	3	75	13	78	6	84	4	78
Nome	−16	78	−6	81	10	87	1	81	3	81
Arizona:										
Phoenix	11	49	19	36	32	37	21	43	21	42
Tucson	10	47	18	26	30	41	21	38	19	38
Yuma	13	43	22	33	34	40	26	40	23	39
Arkansas:										
Fort Smith	4	68	17	63	28	68	17	68	7	67
Little Rock	5	73	17	66	28	71	18	69	7	70
California:										
Fresno	8	80	16	57	27	41	18	54	17	61
Los Angeles	12	59	15	64	21	66	18	64	17	62
Red Bluff	7	72	15	51	27	32	18	47	17	51
Sacramento	7	82	14	66	23	54	17	60	16	66

San Diego	12	68	15	66	20	75	18	72	17	71
San Francisco	9	79	13	77	17	79	16	76	13	77
Colorado:										
Alamosa	−8	68	5	46	18	57	7	56	5	57
Denver	−2	53	8	52	23	48	11	47	10	50
Grand Junction	−3	69	11	43	26	34	13	46	12	48
Connecticut:										
Bridgeport	−2	64	9	63	23	72	13	70	11	70
Hartford	−2	69	8	66	23	73	12	73	10	71
Delaware,										
Wilmington	1	70	11	65	24	71	14	73	12	70
D.C., Washington	3	66	13	59	26	71	15	71	14	67
Florida:										
Jacksonville	13	76	21	70	28	77	22	79	20	75
Miami	19	75	23	72	28	78	26	79	24	76
Orlando	16	75	22	69	28	77	24	75	22	74
Tallahassee	12	77	20	72	27	80	21	75	20	75
Tampa	16	77	22	71	28	78	24	71	22	76
Georgia:										
Atlanta	7	72	16	64	26	73	17	68	17	69
Macon	9	71	18	66	27	76	18	74	18	71
Savannah	11	72	19	68	27	78	20	76	19	73
Hawaii:										
Hilo	22	80	22	81	24	80	24	81	23	81
Honolulu	73	74	23	69	26	67	26	69	24	70
Kahului	22	79	23	75	26	71	25	73	24	74
Lihue	21	80	22	73	24	75	24	77	23	77
Idaho:										
Boise	−2	78	10	56	24	38	11	55	11	58
Lewiston	−1	75	10	52	23	35	11	61	11	57
Pocatello	−3	78	8	54	21	37	9	55	8	53

See footnotes at end of table.

TABLE 28.—*Climatological data for 183 weather stations in the United States, its territories, Puerto Rico, Canada, and Mexico*[1]—Continued

Location of weather station	Average temperature and relative humidity (RH)[2] for—									
	January		April		July		October		Annual	
	Temp.	RH	Temp.	RH	Temp.	RH	Temp.	RH	Temp.	RH
	°C	*Percent*	*°C*	*Percent*	*°C*	*Percent*	*°C*	*Percent*	*°C*	*Percent*
United States—Con.										
Illinois:										
Chicago	−3	76	9	68	24	68	13	69	9	76
Peoria	−4	73	11	66	24	72	13	71	11	72
Springfield	−3	80	12	70	25	69	13	70	12	73
Indiana:										
Evansville	1	76	13	65	26	68	16	69	14	70
Fort Wayne	−3	81	9	69	23	67	12	73	10	73
Indianapolis	−2	74	11	69	24	71	13	71	12	73
Iowa:										
Des Moines	−6	73	10	65	26	69	12	65	10	71
Dubuque	−7	77	9	64	23	67	11	70	9	72
Sioux City	−7	73	9	64	23	66	11	65	9	69
Kansas:										
Concordia	−4	71	12	61	26	60	13	62	12	65
Dodge City	−2	68	12	60	26	57	13	61	13	62
Topeka	−3	71	12	64	26	65	13	64	13	68
Wichita	0	71	14	62	27	62	16	60	14	65
Kentucky:										
Lexington	1	72	12	63	24	72	14	69	13	69
Louisville	2	73	13	64	26	67	15	68	14	69

Louisiana:										
Lake Charles	11	81	21	78	28	80	21	78	20	78
New Orleans	13	77	20	73	28	76	21	72	21	75
Shreveport	9	75	18	69	28	72	19	70	19	71
Maine:										
Caribou	−12	73	2	71	18	75	6	72	4	74
Portland	−6	72	6	70	20	76	10	78	8	74
Maryland, Baltimore	1	65	12	62	25	70	14	71	13	67
Massachusetts:										
Boston	−1	69	9	63	23	69	13	68	10	67
Worcester	−4	68	7	60	21	68	11	70	8	68
Michigan:										
Alpena	−7	76	4	71	19	72	8	77	6	75
Detroit	−3	76	9	65	23	65	12	71	9	70
Grand Rapids	−4	82	8	67	21	66	14	74	9	73
Lansing	−6	79	7	66	22	70	11	77	8	74
Marquette	−7	74	4	69	19	71	8	71	6	74
Sault Sainte Marie	−9	77	3	72	18	77	8	81	4	77
Minnesota:										
Duluth	−13	75	3	69	18	75	7	76	4	75
International Falls	−16	72	3	65	19	72	6	77	3	72
Minneapolis-St. Paul	−11	73	7	63	23	66	9	67	7	69
Mississippi:										
Jackson	9	76	18	74	28	78	19	77	19	76
Meridian	8	76	18	70	27	76	19	76	18	74

See footnotes at end of table.

TABLE 28.—*Climatological data for 183 weather stations in the United States, its territories, Puerto Rico, Canada, and Mexico*[1]—Continued

Location of weather station	Average temperature and relative humidity (RH)[2] for—									
	January		April		July		October		Annual	
	Temp.	RH	Temp.	RH	Temp.	RH	Temp	RH	Temp.	RH
	°C	*Percent*	*°C*	*Percent*	*°C*	*Percent*	*°C*	*Percent*	*°C*	*Percent*
United States—Con.										
Missouri:										
Columbia	−1	71	13	61	26	67	14	71	13	69
Kansas City	−1	68	13	62	26	63	14	62	13	66
St. Louis	3	72	13	64	26	63	15	65	13	67
Springfield	1	76	13	66	24	70	14	69	13	71
Montana:										
Billings	−6	62	7	56	23	46	10	53	8	56
Havre	−10	77	6	59	21	51	8	60	6	64
Helena	−7	67	7	56	19	48	7	61	7	59
Kalispell	−6	78	6	64	18	53	7	72	6	67
Nebraska:										
Lincoln	−4	74	11	66	27	65	14	63	11	69
North Platte	−4	72	9	63	24	64	11	64	9	67
Scottsbluff	−3	67	8	58	23	60	10	59	9	61
Valentine	−6	73	8	63	23	62	10	58	8	66
Nevada:										
Ely	−4	66	6	47	19	33	8	45	7	48
Las Vegas	6	47	18	23	32	22	19	25	19	28
Reno	−1	69	9	49	20	38	9	53	10	53
Winnemucca	−2	74	8	49	22	29	9	41	9	41

New Hampshire,										
Concord	−6	72	6	65	21	72	9	75	10	71
New Jersey,										
Atlantic City	1	69	9	67	22	77	14	77	12	73
New Mexico:										
Albuquerque	2	57	13	36	26	43	14	45	13	45
Roswell	4	54	14	38	26	51	16	54	15	44
New York:										
Albany	−4	71	8	64	22	70	11	74	9	70
Binghamton	−4	77	7	69	21	75	10	76	8	75
Buffalo	−4	77	7	80	21	69	11	73	8	73
Canton	−9	78	6	69	19	70	8	75	8	74
New York	1	67	11	64	25	68	15	68	12	67
Syracuse	−4	75	7	67	21	69	11	73	9	74
North Carolina:										
Asheville	2	77	13	70	23	84	13	83	13	78
Charlotte	6	67	16	62	26	75	17	72	16	69
Raleigh	6	72	15	64	26	76	16	76	16	72
North Dakota:										
Bismarck	−12	72	7	64	22	64	8	66	5	68
Devils Lake	−16	75	4	67	19	69	6	70	3	71
Fargo	−14	74	6	69	22	68	8	70	4	71
Williston	−12	71	6	60	22	58	8	64	4	65
Ohio:										
Cincinnati	2	77	13	67	26	70	16	71	13	71
Cleveland	−3	78	8	67	22	67	12	69	10	71
Columbus	−1	78	11	67	24	68	12	70	11	71
Toledo	−3	73	9	68	23	72	12	72	10	72
Oklahoma:										
Oklahoma City	3	70	15	62	27	64	17	66	16	66
Tulsa	3	67	16	62	28	65	17	67	16	66

See footnotes at end of table.

TABLE 28.—*Climatological data for 183 weather stations in the United States, its territories, Puerto Rico, Canada, and Mexico*[1]—Continued

Location of weather station	Average temperature and relative humidity (RH)[2] for—									
	January		April		July		October		Annual	
	Temp.	RH	Temp.	RH	Temp.	RH	Temp.	RH	Temp.	RH
	°C	*Percent*	°C	*Percent*	°C	*Percent*	°C	*Percent*	°C	*Percent*
United States—Con.										
Oregon:										
Baker	−3	78	7	61	19	54	8	65	7	67
Burns	−3	77	7	54	21	37	8	56	8	58
Eugene	4	88	10	75	19	63	12	83	11	77
Portland	3	83	11	71	19	67	12	81	11	75
Roseburg	5	87	11	68	19	60	7	79	12	74
Pennsylvania:										
Harrisburg	−1	68	11	63	22	67	12	70	12	67
Philadelphia	0	70	11	65	24	70	13	72	13	69
Pittsburgh	−2	73	9	65	22	68	12	68	12	68
Rhode Island,										
Providence	−2	68	8	66	23	74	12	74	11	71
South Carolina:										
Charleston	10	76	18	72	27	82	19	71	19	70
Columbia	9	71	17	64	28	73	18	71	18	70
South Dakota:										
Huron	−12	75	8	65	22	65	9	66	7	70
Rapid City	6	67	7	56	23	54	10	53	8	60
Sioux Falls	9	73	8	66	23	67	11	68	8	70
Tennessee:										

Knoxville	4	74	14	62	25	72	15	71	15	75
Memphis	6	74	17	66	28	71	18	69	17	70
Nashville	4	76	16	64	27	69	17	69	16	70
Texas:										
Abilene	7	62	18	56	28	54	18	61	18	59
Amarillo	3	62	14	53	27	56	16	60	14	58
Brownsville	16	79	23	77	29	76	24	75	23	77
Del Rio	11	66	21	56	29	57	21	65	21	61
El Paso	6	44	17	27	28	43	18	45	18	39
Fort Worth	8	68	18	63	29	59	20	62	19	63
Houston	12	83	21	81	28	77	22	75	20	79
Midland	7	52	18	44	28	47	18	58	17	53
San Antonio	11	69	20	66	29	64	22	66	21	67
Utah:										
Milford	−3	63	9	31	24	22	10	32	9	38
Salt Lake City	−3	76	10	53	24	38	12	54	11	56
Vermont,										
Burlington	−8	76	6	65	22	70	9	74	7	73
Virginia:										
Norfolk	5	72	14	68	26	77	17	75	16	70
Richmond	3	73	14	64	25	75	15	77	14	72
Roanoke	3	69	14	59	24	71	14	68	13	66
Washington:										
Seattle	3	80	9	69	18	66	11	81	11	74
Spokane	4	81	8	58	21	44	9	68	9	64
WallaWalla	1	79	12	54	24	39	12	61	12	60
Yakima	−2	80	11	51	22	47	11	67	11	62
West Virginia:										
Charleston	3	69	13	60	24	77	14	73	13	70
Elkins	−1	77	9	69	21	82	12	78	10	77
Huntington	2	71	13	61	24	77	14	72	13	71
Parkersburg	1	74	12	62	24	72	13	72	12	71

See footnotes at end of table.

TABLE 28.—*Climatological data for 183 weather stations in the United States, its territories, Puerto Rico, Canada, and Mexico*[1]—Continued

Location of weather station	Average temperature and relative humidity (RH)[2] for—									
	January		April		July		October		Annual	
	Temp.	RH	Temp.	RH	Temp.	RH	Temp.	RH	Temp.	RH
	°C	*Percent*	*°C*	*Percent*	*°C*	*Percent*	*°C*	*Percent*	*°C*	*Percent*
United States—Con.										
Wisconsin:										
Green Bay	17	73	6	67	22	68	9	72	7	71
LaCrosse	16	72	8	67	23	74	11	74	8	73
Madison	17	76	8	62	22	67	10	71	8	72
Milwaukee	−7	74	6	71	21	70	11	72	8	72
Wyoming:										
Casper	−5	63	6	59	22	46	8	56	7	57
Cheyenne	−3	55	5	59	19	53	8	54	7	56
Lander	−7	62	6	53	21	41	7	53	7	53
Sheridan	−7	67	6	60	19	52	7	57	7	61
American Samoa, Pago Pago	27	84	27	86	26	82	27	84	27	83
Guam	25	83	26	81	26	87	26	87	26	84
Puerto Rico, San Juan	23	78	25	78	27	81	27	82	27	80
Canada:										
Alberta:										
Edmonton	−14	76	4	58	17	63	5	61	3	67
Lethbridge	−8	71	6	55	19	50	7	53	6	59
British Columbia, Penticton	−3	81	9	57	20	49	9	69	9	65

Manitoba, Winnipeg	−18	78	3	68	20	64	6	69	3	70
Nova Scotia,										
Halifax	−5	83	4	75	19	80	9	82	7	79
Ontario:										
London	−5	84	6	73	21	69	10	78	8	77
Ottawa	−11	75	6	60	21	66	8	71	6	69
Quebec:										
Montreal	−9	75	6	63	21	67	9	70	7	70
Quebec	−12	76	3	64	19	70	7	74	4	71
Saskatchewan:										
Regina	−17	81	3	63	19	59	5	65	2	69
Saskatoon	−17	76	4	62	19	59	5	63	2	67
Mexico:										
Chihuahua	11	50	19	36	26	52	18	51	18	46
Ensenada	13	79	15	81	20	82	18	82	17	80
Guaymas	19	52	24	48	32	63	31	67	15	78
Mexico City	11	53	18	45	18	67	16	65	16	58
Monterrey	14	58	23	59	27	62	22	69	22	63
Soto la Marina	18	78	25	75	28	76	24	79	24	77
Tampico	19	78	24	78	28	79	27	77	24	78
Torreón	16	57	23	47	27	51	22	58	22	53

[1] Data from (1) "Local Climatological Data, Annual Summary With Comparative Data," 1972, and (2) "Climatic Atlas of the United States," 1968, both published by National Oceanic and Atmospheric Administration, U.S. Department of Commerce; (3) "Atlas of American Agriculture," U.S. Department of Commerce, 1928; (4) "U.S. Naval Weather Service World-Wide Airfield Summaries," v. VII (Central America), February 1968; and (5) Meteorological Branch, Canada Department of Transport.

[2] Each average relative humidity, except one, represents the average of 4 readings taken at approximately 12 p.m., 6 a.m., 12 m., and 6 p.m. For Milford, Utah, the average represents readings taken at 11 a.m. and 5 p.m., the only data available. Each fraction resulting from averaging temperatures and relative humidities is rounded to the nearest whole number; 0.5 is raised to the next number.

Using the correct seed moisture content for sealed storage is very critical as seed sealed with too high a moisture content may lose viability more rapidly than air-dry seed in open storage at the same temperature. For very long-term storage, seed should be dried to 5-percent or lower moisture content, sealed in moisture-barrier containers, and held at −17.8° C, whereas storage at 4° and 50-percent relative humidity in open containers is satisfactory for many kinds of seed for 10 years or longer.

There are, of course, exceptions, as some kinds of seeds cannot endure freezing and others cannot tolerate dehydration without losing viability. The cost of storing a given quantity of seed increases rapidly as the storage conditions become more exacting. Thus, the storage facility should be related to storage time, commensurate with the needs for maintaining satisfactory levels of seed viability and vigor. All storage structures must be designed to protect the seeds from insect and rodent infestations.

It should be emphasized that the intended storage period is only one of several items to be considered when deciding on storage conditions for a particular lot of seed. The more significant of these are seed factors, storage environment, effect of pests, and storage structures.

Care of Seeds in Transit

Storage Principles Applicable to In-Transit Seeds

As pointed out by Harrington *(1972)* and others, seeds are in storage from the date of their physiological maturity until they are planted. Thus, seeds in transit are, in fact, in storage. As used here, the expression "in transit" includes not only the time the seeds are being moved from one location to another but also the time they are awaiting shipment, whether in a warehouse, railroad car, truck, airplane, boat, or on a dock awaiting further consignment. While in transit the seeds are subject to the same storage principles as seeds in warehouses. The principal differences between warehouse storage and in-transit storage are the relatively fast changes in the seed environment that can result from modern transportation and the failure of personnel to foresee or predict all the hazardous conditions to which the shipment may be subjected. The two principal envionmental factors that determine whether a seed shipment will arrive safely at its destination are the temperatures to which the seeds are exposed and the seed moisture content, which is controlled by the relative humidity of the air or protective packaging.

Historical Background

Moodie *(1925)* reported on the effect of temperature on seeds shipped from Australia to England and from England to Australia. Since the temperatures used aboard ship were not precise and moisture content of the seeds was not controlled, his results are not completely convincing. From his experiments Moodie concluded that seeds stored under cool conditions with little variation are likely to retain viability better than when stored under warm conditions with extensive variations in temperature. von Degen and Puttemans *(1931)* shipped predried and nondried seeds of several species of field crops and vegetables between Hungary and Brazil. Their results proved conclusively the beneficial effect on germination of predrying the seeds before shipment and maintaining low moisture content during transit.

Foy *(1934)*, working at Palmerston North, New Zealand, and Kearns and Toole *(1939)*, working at Washington, D.C., proved conclusively through cooperative studies that Chewings fescue seed could be shipped between New Zealand and the United States and between New Zealand and England without serious loss of viability by predrying the seed before shipment and avoiding exposure to high temperatures and humidities. Their carefully conducted studies, which included seeds exchanged between the three countries and seed stored at Washington, Palmerston North, and Cambridge, England, were carried out on only one kind of seed, Chewings fescue. However, the basic principles relating to seed moisture content as well as relative humidity and temperature during shipment govern the maintenance of viability of transoceanic shipment of various kinds of seeds. The results are so convincing that very little research of this nature has been reported since.

The deleterious effects of high temperature, high relative humidity, or both are mitigated by the length of time the seed will be under a particular set of conditions. For example, a shipment of seed from Cape Town, South Africa, direct to New York by modern freighter may arrive at its destination without any loss in germination. On the other hand, a similar shipment by tramp freighter, which would spend up to 6 weeks stopping at Belem and Manaus, Brazil, might arrive at New York with significantly reduced germination. Both cargoes in this hypothetical example passed through tropical waters; however, the extra time the seed was exposed to hot humid conditions in the Tropics could be disastrous for the seed.

Some Hazards To Be Avoided

A person planning to ship seeds should anticipate the various hazards to seeds during transit and while in storage immediately before and after shipment. These hazards will vary depending on the method of transportation—whether by truck, railroad car, ship, or airplane. Rapid changes of temperature or relative humidity resulting from the movement of seeds between zones of different temperatures and humidities can create hidden problems. Seeds chilled during air transportation are likely to be covered with condensed water after landing of the plane at warmer temperatures. Under some conditions this surface moisture can be absorbed by the seeds rather rapidly. Proper storage immediately after landing will obviate this problem.

Ching *(1959)* pointed out that under simulated shipping and storage experiments Highland bentgrass seed absorbed water rapidly under tropical conditions and lost its viability within a few weeks. It is noteworthy that Ching found seed stored under arctic conditions tended to absorb water gradually during prolonged storage and consequently gradually lost viability. She recommended that shippers not send high moisture seed, such as that stored at high humidities, during the winter into areas where it would be exposed to high temperatures.

Harrington *(1972)* pointed out the peril of seeds becoming overheated when shipped in boxcars placed on a railroad siding in a warm or hot climate for several days. The boxcar acts as a heat trap whereby very high temperatures may be reached. This hazard is greatest with high moisture seeds.

Shipping seeds with borderline to high moisture contents is no doubt the greatest of all hazards. Christensen *(1969)* stated that much grain is shipped and stored at moisture contents slightly above safe levels at anticipated temperatures. When the temperature rises purposely, accidently, or unexpectedly, respiration increases, storage molds appear, and spoilage results. This is more likely to occur in grain shipped in bulk than in seeds ordinarily shipped in bags or smaller containers.

Still another hazard is the possibility of shipping seeds as a part of a mixed cargo. When considering this possibility, the shipper should be sure that no chemicals that are deleterious to the seeds, such as certain herbicides, are included in the cargo. Any substance that might increase the concentration of oxygen should be excluded as oxygen hastens seed deterioration. There is no objection to inclusion of the inert gases, such as hydrogen, nitrogen, and carbon dioxide, which tend to retard deterioration.

Another hazard is the possibility of the seedsman or shipper failing to carefully select the seed lots. Seeds of different crop species vary as to their storability. Oily seeds deteriorate more rapidly than starchy seeds. Storability differences may exist among cultivars of some crop species and also among different crops of the same cultivars or species. Such information when available should be used in choosing seed lots for shipment. If seeds are to be shipped long distances under unfavorable or questionable conditions, only fresh seeds of high viability and vigor should be chosen.

General Recommendations

(1) Carefully select the seed lot for shipment. Consider the general guidelines given previously.

(2) Carefully consider such factors as the shipping quality of the seed lot selected, method of shipment, distance to be shipped, time in transit, and climatic zones through which the seed will pass.

(3) If the seed is not of safe moisture content for shipping, reduce moisture to a safe level by an acceptable drying method.

(4) If feasible, ship the seed under controlled temperature and relative humidity to assure viability. If not feasible, reduce the seed moisture content to a safe level and ship in moisture-resistant or moistureproof containers (Caldwell and Bunch, *1961*). See the sections on "Seed Storage Structures" and "Packaging and Packaging Materials."

THEORIES REGARDING SEED DETERIORATION

We have pointed out how growing, harvesting, processing, and storage conditions affect seed longevity. We have also discussed ways to slow down the rate of seed deterioration, but we have not discussed what occurs inside a dry, sound, disease-free seed to cause it to weaken and die. Several theories based on genetic and physiological principles have been proposed.

Changes in Protein Structure

Ewart *(1908)* theorized that seed longevity depends not on available food reserves but on how long the proteid molecules, into which protoplasm disintegrates when drying, can recombine into active protoplasm with the absorption of water. According to this reasoning, protein molecules should disintegrate excessively when seeds are dried to very low moisture levels. Struve,[10] however, concluded that corn dried to less than 4-percent moisture, sealed in nitrogen, and stored at a low temperature would keep indefinitely. Nutile *(1964b)* found that drying vegetable seeds to 0.4 percent caused some damage to carrot, celery, eggplant, pepper, tomato, and Kentucky bluegrass seeds but not complete loss of viability. Seeds of cabbage, cucumber, lettuce, onion, and Highland bentgrass sealed with 0.4-percent moisture were not injured.

Crocker *(1938)* suggested that protein coagulation caused loss of viability. Later he *(1948)* reported that his protein coagulation theory was too general because of the many kinds of protein in embryos and because his studies did not show which protein coagulates with aging.

Depletion of Food Reserves

The theory that depletion of food reserves for the embryo causes

[10] See footnote 8, p. 61.

seeds to die did not persist long because it was soon evident that many dead seeds still contained ample food reserves. Some *Zea mays* L. seeds, more than 700 years old, found in the Mesa Verde cliff dwellings, appeared sound visually, yet not a single viable seed was ever found among them. Oxley *(1948)* suggested that exhaustion of an unnamed organic compound results in loss of seed viability. Harrington *(1960b)* reasoned that a seed may have an adequate food supply and still die because of a breakdown in the food transport system. Seed moisture content may be high enough for respiration but too low to transport food from the reserve supply to the embryo (Harrington, *1967*). Although food reserves are usually not entirely depleted when seeds die, various changes may have occurred, such as increased acidity (Zeleny and Coleman, *1938, 1939*; Milner and Geddes, *1946*) and decreased lipids and proteins (Pomeranz, *1966*; Ching and Schoolcraft, *1968*; Koostra and Harrington, *1969*).

Development of Fat Acidity

The development of fat acidity in seeds has been shown to accompany death. Germination declined 8 percent and fat acidity increased 14 units when 8- to 9-percent moisture content soybeans were stored for 700 days. With increased seed moisture content, germination dropped rapidly and fat acidity increased sharply (Holman and Carter, *1952*). A drop in the germination of wheat seeds was accompanied by an increase in fat acidity (Kelly et al., *1942*). Increased fat acidity was also a major cause of corn seed deterioration (Zeleny and Coleman, *1939*). Fat acidity of peanuts increased only after the stored seeds were dead (Davis, *1961*). Although fat acidity values have been established for grain showing little or no deterioration (Baker et al., *1957*), such values may not constitute a reliable index of viability (James, *1967*).

Enzymatic Activity

Attempts have been made to use enzymatic activity as a measure of seed viability. However, only a few of the many enzymes in seeds have been investigated. Early work with enzymes dealt principally with catalase activity. Crocker and Harrington *(1918)* found catalase activity in dead Johnsongrass and yet reported a relationship with viability. They could not establish such a relationship with seeds of *Amaranthus retroflexus*. Davis *(1926)* showed a relationship between catalase ratio and viability of lettuce seeds, but the range of his study was not wide enough to establish a linear relationship. Leggatt *(1929–30)* obtained a correlation between catalase activity and germination of wheat seeds. Because of the limited number of species studied and the inconsistent results, the relationship between catalase activity and seed viability appears questionable.

Phenolase activity also does not appear to be an indicator of viability (Davis, *1931*). For wheat, Davis found a relationship with germination but not age, and for oats a relationship with age but not germination. Throneberry and Smith *(1955)* reported that malic dehydrogenase activity was more closely correlated with germination percentage than was alcohol dehydrogenase and cytochrome dehydrogenase activity. They believed inactivation of these enzymes was not a major cause of viability loss. Linko and Sogn *(1960)*, Bautista and Linko *(1962)*, and Grabe *(1964)* demonstrated a close relationship between glutamic acid decarboxylase activity and viability. James *(1968)* found that the correlation coefficient between germination percentage and glutamic acid decarboxylase activity for the average of all bean varieties studied compared favorably with the coefficients reported by Grabe *(1964)*; for corn, the coefficients for individual varieties were variable. James *(1968)* stated that the inconsistencies of the correlation coefficients among the cultivars used definitely limit glutamic acid decarboxylase activity as an indicator of viability of bean seeds. Because of the conflicting reports on the relationship of enzymatic activity to seed deterioration, much research remains to be done in this area.

Chromosomal Changes

In 1901 De Vries (*in* Kostoff, *1935*) observed that old *Oenothera lamarckiana* seeds produced more abnormal plants than did fresh seed. More recently chromosomal changes have been reported in old seeds of a relatively large number of species: *Crepis* spp. (Navashin, *1933*; Navashin and Shkvarnikov, *1933*; Navashin and Gerassimowa, *1936*), corn (Peto, *1933*), onion (Nichols, *1941, 1942*), sugar beet (Lynes, *1945*), barley, pea, rye, and wheat (Gunthardt et al., *1953*), *Datura* spp. (Avery and Blakeslee, *1936*; Blakeslee et al., *1942*; Blakeslee, *1954*), and barley, broadbean, and pea (Abdalla and Roberts, *1969a*).

Although numerous compounds are known to induce mutations, few of them are normally found in seeds. Mutagenic compounds found in plants include adenine, degradation products of adenine, uracil, thymine, adenosine desoxyribonucleic acid, and ribonucleic acid (D'Amato and Hoffman-Ostenhof, *1956*). The degrees of mutagenic action given by these authors are (1) the lethal zone, where the accumulation of mutagens becomes toxic and kills the seed, (2) the narcotic zone, which affects the spindle mechanism, and (3) the subnarcotic zone, in which the mutations occur. The mutagenic or chromosomal aberration theory is supported further by the fact that (1) extracts from old seeds induce mutations in fresh seeds, (2) the mutation rate increases with age, (3) spontaneous mutations in dormant seeds become evident in the chromosome presplit phase and in adult plants, and (4) reactions of root and shoot tips in old seeds closely parallel the reactions of the same kind of seeds when treated with X-ray. Apparently mutagens do not develop under good storage conditions, as all observations have been on seeds affected by high temperature, relative humidity, or both.

Blakeslee's *(1954)* work with *Datura* demonstrated that age alone was not responsible for the development of mutations. He found that mutations developed in seeds at room temperature. The mutation rate for seeds buried in the ground for 39 years was extremely low. James *(1967)* suggested that seeds presently housed in the National Seed Storage Laboratory at Fort Collins, Colo., may hold the ultimate answer to the question of the relationship of mutations to seed age. However, the answer will not be determined soon because the storage conditions there are such that respiration in the stored seeds appears to be near the minimum.

Membrane Damage

According to Villiers *(1973)*, the immediate damage rendering aged seeds incapable of germination is extranuclear. Free radical damage to membranes and enzyme systems could affect essential metabolic processes when the seeds become imbibed for germination. Harrington *(1973)* stated that seeds dried to 4- to 5-percent moisture content appear to deteriorate slightly faster than seeds with 5- to 6-percent moisture, probably from lipid autoxidation damage. Unsaturated lipids in seed cells may break, producing two free radicals, which can react with other lipids, destroying the structure of cell membranes. In imbibed cells tocopherols made by enzymes combine with free radicals rendering them harmless. Since enzymes are inactive at low seed moisture contents, the free radicals produced nonenzymically become destructive when the tocopherols that were present when the seeds were dried are depleted.

Respiration

The theories for seed deterioration, except possibly fat acidity, are related to respiration. Respiration increases in proportion to the amount of moisture in seeds, but it is very low at moisture contents between 4 and 11 percent (Baily, *1940*; Harrington, *1963*). Respiration rates up to about 50° C are also directly proportional to temperature. With high temperatures and high moisture contents, seeds lose viability very rapidly, usually in less than 3 months at 32° and 90-percent relative humidity. Seeds stored at temperatures below 10° and at low relative humidities will remain viable for a long time. According to James *(1967)*, peanut seeds recognized as short lived were held at the Southern Regional Plant Introduction Station, Experiment, Ga., for 8 years at 10° and 50-percent relative humidity without a significant decrease in viability.

Summary

What really causes a seed to deteriorate? We examined the theory of changes in protein structure and found that no specific changes have been pinpointed. Depletion of food reserves is not a very sound theory as almost all seeds contain an abundance of food long after the embryo is dead. Although the development of fat acidity has been shown to accompany the death of a seed, such development has not been firmly estalished as the cause of death. No specific enzyme activity change has been proved to cause the death of seeds. Chromosomal changes have been noted in various kinds of seeds; however, no one has yet proved conclusively whether such changes are really the cause of deterioration or are just a result. Membrane damage, though not specifically established as the cause of deterioration, is definitely associated with it.

Although several theories have been proposed, none satisfactorily explains how seeds deteriorate. Even though the process of deterioration is not clearly understood, the methods for preventing it are well established. For more detailed information on theories of seed deterioration, see Streeter *(1965)*, Roberts *(1972)*, Abdul-Baki and Anderson *(1972)*, and Heydecker *(1973)*.

OLD AND ANCIENT SEEDS

Ewart *(1908)* wrote as follows about the longevity of seeds:

"Probably few sections of human knowledge contain a larger percentage of contradictory, incorrect and misleading observations than prevail in the works dealing with this subject, and, although such fables as the supposed germination of mummy wheat have long since been exploded equally erroneous records are still current in botanical physiology. In addition, there are considerable differences of opinion as to the causes which determine the longevity of seeds in the soil or air. The works of de Candolle 1832, 1846, de Candolle and Picket 1895, Duvel 1905 and Becquerel 1934 are the most accurate and comprehensive dealing with the question. The subject is still, however, in an incomplete and fragmentary condition."

The principal objectives here are twofold. First, we establish that very old seeds in some instances have retained vitality over a long period, even more than a century. We differentiate (1) crop seeds known to have survived for relatively long storage periods, (2) seeds of native plants known to have survived longer than crop seeds, and (3) circumstantial evidence of seed survival for a few to several centuries.

The second principal objective is to analyze the authenticity of so-called mummified seeds or grains.

Maximum Known Survival of Crop Seeds

Unless crop seeds are kept under favorable storage conditions, they lose viability within a few years. The information in the section on "Examples of Storability of Different Plant Species" confirms this broad-based general statement. On the other hand, there are several authentic records of cereal crop seeds surviving up to 32 years. Aufhammer and Simon *(1957)* reported survival of barley and oat seeds for 123 years (table 29). Authentic records show that seeds of some of the leguminous crops have survived up to 81 years (table 29).

All the results in table 29 are regarded as authentic. Some data have been obtained from planned experiments and others from tests on seeds stored in museums, cupboards, or laboratories under such conditions that the history of the seed was known.

Maximum Known Survival of Primarily Weed and Native Plant Seeds

Mature seeds of many weeds and native plants have dormant embryos, a condition that retards germination. In fact, seeds do not germinate until they undergo physiological and biochemical changes that permit afterripening. A dormant or quiescent embryo appears to enhance longevity. Also, seeds with hard coats, which restrict imbibition of water and exchange of gases, are among those with the greater lifespans. Many seeds of weeds and native plants are in this group, which includes such plant families as the Convolvulaceae, Leguminosae, Malvaceae, and Nymphaeaceae.

Plant species whose seeds have long lifespans are listed in table 30. The data were obtained from seeds kept or stored dry with known histories and from seeds buried in soil by planned experimentation.

Seeds Stored Dry

Perhaps the earliest authentic records of tests on the continued vitality of air-dry seeds are those of Alphonse de Candolle *(1846)*. He *(1832)* tested seeds of different species harvested in 1831 and kept them all air-dry until May 1846 (15 years), when he planted 20 seeds of each species. There were 368 species representing 53 families. Five out of 10 species of Malvaceae, 9 out of 45 species of Leguminosae, and 1 out of 30 species of Labiatae showed some germination.

Ewart *(1908)* published the results of tests of over 1,000 species of seeds, which he found locked in a cupboard in the botanical laboratory at Melbourne, Australia. The seeds had been sent from Kew Gardens,

TABLE 29.—*Examples of documented long lifespans of some common crop seeds when maintained under favorable survival conditions*

Crop	Storage condition	Place stored	Storage period	Germination	Reference
			Years	*Percent*	
Barley	Ambient temperature and relative humidity.	Eastern Washington; dry climate.	32	96	Haferkamp et al. *(1953)*.
	In glass vial embedded in foundation stone of building.	Nuremburg, Germany	123	12	Aufhammer and Simon *(1957)*.
Corn	At low relative humidity.	Fort Collins, Colo.	21	32	Robertson et al. *(1943)*.
	Ambient temperature and relative humidity.	Eastern Washington; dry climate.	32	11, 19, 23, 53, 70 (for different cultivars).	Haferkamp et al. *(1953)*.
Oats	do	do	32	84	Do.
	In glass vial embedded in foundation stone of building.	Nuremberg, Germany	123	22	Aufhammer and Simon *(1957)*.
Sorghum	In laboratory, dry	Fort Collins, Colo.	17	98	Robertson et al. *(1943)*.
	In envelopes in laboratory.	Chillicothe, Tex.	19	.5	Karper and Jones *(1936)*.
Wheat	Ambient temperature and relative humidity.	Eastern Washington; dry climate.	32	85	Haferkamp et al. *(1953)*.
	In laboratory, dry	Fort Collins, Colo.	26	19.5	Fifield and Robertson *(1952)*.

TABLE 29.—*Examples of documented long lifespans of some common crop seeds when maintained under favorable survival conditions*—Continued

Crop	Storage condition	Place stored	Storage period	Germination	Reference
Wheat—Con.			*Years*	*Percent*	
	Ambient storage; dry climate.	Australia	28	60	Brown and Myers *(1960)*.
	Approx. 4.5- to 4.8-percent moisture content.	England	32	69	Whymper and Bradley *(1947)*.
Bluegrass, Kentucky.	In soil; buried seed experiment.	Arlington, Va.	39	1-2	Toole and Brown *(1946)*.
Alfalfa	Ambient temperature and relative humidity.	Eastern Washington; dry climate.	33	49	Haferkamp et al. *(1953)*.
Clover, red	In loosely corked bottle in museum.	London, England	81	2.6	Turner *(1933)*.
	Same seed samples	do	100	1	Youngman *(1952)*.
	In soil; buried seed experiment.	Arlington, Va.	39	2, 16	Toole and Brown *(1946)*.
Pea (garden and field):					
Bluebell	Ambient temperature and relative humidity.	Eastern Washington; dry climate.	31	46	Haferkamp et al. *(1953)*.
Canada	do	do	30	39	Do.
Smilen	do	do	30	44	Do.
Solo	do	do	31	78	Do.

Sweetclover, white.	In loosely corked bottle in museum.	London, England	81	.6	Turner *(1963)*.
Trefoil, big.	do	do	81	9.6	Do.
	Same seed samples	do	100	1	Youngman *(1952)*.
Beet, sugar	In metal containers for 10 years, then in cloth sacks; below freezing.	Salt Lake City, Utah	22	75	Pack and Owen *(1950)*.
Tobacco	In sealed containers, desiccated, and refrigerated.	Quincy, Fla.	25	High viability.	Kincaid *(1958)*.
	In soil; buried seed experiment.	Arlington, Va.	39	Approx. 20	Toole and Brown *(1946)*.
Cotton	In laboratory	Knoxville, Tenn.	25	6	Simpson *(1946)*.
Carrot	do	Cheyenne, Wyo.	31	7	James et al. *(1964)*.
Cucumber	do	do	30	77	Do.
Muskmelon	do	do	30	96	Do.
Onion	do	do	22	33	Do.
Parsnip	do	do	28	30	Do.
Pepper, red	do	do	28	69	Do.
Watermelon	do	do	30	92	Do.
Strawberry	At approx. −12° C	Beltsville, Md.	23	89	Scott and Draper *(1970)*.

TABLE 30.—*Records of seeds 75 years or older with some germination*

Reference	Source and condition	Species	Age when tested	Germination
			Years	*Percent*
Ewart *(1908)*	Seeds received from Kew Gardens, London, and stored in a locked cupboard in Melbourne, Australia.	*Aleurites moluccana*	79	74
		Cytisus candicans	80	62
		Goodia latifolia	105	8
		Hovea linearis	105	17
		Indigofera cytisoides	81	5
		Melilotus alba	77	18
		Melilotus gracilis	82	22
Becquerel *(1907, 1934)*.	Storage room in National Museum, Paris; *Cassia multijuga* seed collected in 1776, obtained from another source; 10 seeds used in each test except 2 for *C. multijuga*.	*Astragalus massiliensis*	86	10
		Cassia bicapsularis	115	40
		Cassia multijuga	158	100
		Dioclea pauciflora	93	20
		Leucaena leucocephala	99	30
		Mimosa glomerata	81	50
Turner *(1933)*	From museums in England where seeds were preserved in loosely corked bottles.	*Anthyllis vulneraria*	90	4
		Cytisus scoparius	81	.6
		Lotus uliginosus	81	9.6
		Medicago orbicularis	78	22
		Melilotus alba	81	.6
		Trifolium pratense	81	2.6
		Trifolium striatum	90	14.1
Schjelderup-Ebbe *(1936)*.	Stored dry or in bottles or paper bags in Norway.	*Astragalus utriger*	82	6
		Kennedya apetala	77	13

Anonymous *(1942b)*	Stored in British Museum, London	*Albizia julibrissin*	147	Some seeds germinated.
	From Hans Sloane collection in British Museum.	*Nelumbo nucifera*	150	2 seeds tested; both germinated.
			237	1 seed tested; it germinated.
Darlington and Steinbauer *(1960)*.	Beal's buried seed experiment at East Lansing, Mich.	*Oenothera biennis*	80	10
		Rumex crispus	80	2
		Verbascum blattaria	80	70
Kivilaan and Bandurski *(1973)*.	do	*Verbascum blattaria*	90	20

England, in 1856 for the University Gardens, but they had been put away in a dark, dry closet, where they had remained unopened for 50 years or longer. He soaked the seeds and placed them on moist filter paper in glass dishes in germinators. Seeds that did not swell after 1 or 2 days in water were either filed or treated with concentrated sulfuric acid. After washing, the seeds swelled readily. Out of 1,400 kinds tested, Ewart *(1908)* found 58 that retained their vitality after 50 years or more in storage. He also tested seeds from another source that varied in age up to 105 years. Of these seeds, *Goodia latifolia* and *Hovea linearis* germinated 8 and 17 percent, respectively. Of the 58 kinds surviving 50 years or more, 36 were in the Leguminosae, 4 in the Malvaceae, 2 in the Euphorbiaceae, and the remainder in miscellaneous plant families. Dry seeds of many leguminous species are frequently hard. The results of some of Ewart's tests are in table 30.

Becquerel *(1934)* had access to a batch of old seeds in a storage room in the National Museum of Paris. These seeds were collected from 1819 to 1853. He conducted germination tests on them in 1906 and again in 1934. Since they were all hard-coated seeds, they required special treatment. They were sterilized, the coats were broken, and they were put in tubes under sterile conditions at 28° C to germinate. The seed stock was considered so precious that only 10 seeds of each kind were used for the test. For the 1934 test he obtained from another source about 20 seeds of *Cassia multijuga*, which were collected in 1776. Only two of these seeds were tested. Table 30 shows the results obtained with six species that germinated after 75 or more years. All these seeds were Leguminosae. The seeds of *C. multijuga* germinated after 158 years of storage. Becquerel believed the long lifespan of all these seeds was made possible by impermeability of the coats, which prevented any exchange of gases or water between the interior of the seed and the outside atmosphere, and by the high degree of desiccation (2- to 5-percent moisture) and absence of oxygen in which the embryos existed within the hard coats.

Turner *(1933)* tested the viability of old seeds from different sources kept in loosely corked bottles in a museum. Of nine species listed in his publication, seven retained viability in excess of 75 years (table 30). All were Leguminosae.

Schjelderup-Ebbe *(1936)* tested the viability of 1,254 batches (nearly as many species) of seeds stored in bottles or paper bags for 34 to 112 years. Seeds of 3 species survived for 50 to 59 years, 10 for 60 to 69 years, 3 for 70 to 74 years, and 2 for over 75 years (table 30). Of 54 entries in the author's table, 19 species germinated after 50 or more years of storage. One species belonged to the Cannaceae, 13 to the Leguminosae, 4 to the Malvaceae, and 1 to the Convolvulaceae. Seeds of some species in all these plant families produce hard seeds.

Buried Seed Experiments

The first planned experiment in America to test the longevity of seeds buried in soil was begun in Michigan. In the fall of 1879, Beal *(1885)* of the Michigan Agricultural College began an experiment to determine the longevity of seeds of 23 kinds of weeds growing near East Lansing, Mich. Fifty freshly harvested seeds of each of the 23 kinds (2 woody and 21 herbaceous) were mixed with sand and placed in pint bottles. The 20 bottles were buried. They were inclined at a 45° angle 18 inches below the soil surface on a sandy knoll of the college campus.

The plan was to dig up one bottle every 5 years and test the seeds for germination. This schedule was followed until 1920, when it was decided to remove bottles at 10-year intervals to extend the experiment. Seeds of *Brassica nigra* and *Polygonum hydropiper* showed some germination for the last time in the 50th year (Darlington, *1931*). Seeds of *Oenothera biennis, Rumex crispus,* and *Verbascum blattaria* germinated in each succeeding test until the 80th year (Darlington and Steinbauer, *1960*). As indicated in table 30 for Beal's experiment, *V. blattaria* was the only species that germinated in the 90th year (Kivilaan and Bandurski, *1973*).

Duvel *(1905)* set up a more elaborate buried seed experiment in 1902 at Arlington, Va. It was designed to test the longevity of 107 species of crop and weed seeds. The seeds were mixed with sterilized soil, placed in porous flowerpots, covered with inverted flowerpot saucers, and buried at depths of 8, 22, and 42 inches. After 20 years' burial some seeds of 51 of the 107 species were still alive. They were primarily weed seeds and hard-seeded legume species. Of 31 cultivated species in the experiment, 22 were dead after 20 years in the soil and most of them after 1 year. The following cultivated species showed some germination after 20 years: Alsike clover, beet, bushclover, celery, Kentucky bluegrass, red clover, timothy, tobacco, and white clover. Forty-four species grew after 30 years and 36 species after 39 years. The experiment was discontinued in 1941 (after 39 years) just before the Arlington Experimental Farm of the U.S. Department of Agriculture was occupied by the Defense Department.

A seed longevity experiment was started by sealing 20 lots of 120 kinds of seeds in 2,400 glass tubes (Went, *1948*). The seeds were dried before they were stored in vacuum tubes. One lot of each kind of seed was to be removed at 10-year intervals at first and later at 20-year intervals until A.D. 2307. The object of this experiment was to obtain data on (1) how long seeds can be stored without losing their power to germinate, (2) how they are affected by long storage, and (3) what evolutionary changes occur from year to year in plants of the same species.

Circumstantial Evidence of Long Lifespans of Seeds

A large number of presumed long lifespans of seeds can be cited. The age of the seeds has been estimated by such comparisons as with objects of known age, surroundings, geological events, and carbon-14 dating. In some cases, the seeds have been tested by reliable procedures, but in other instances, germination in nature has been observed and accepted as fact. Observations of the appearance of seedlings cited as evidence of seed longevity or persistence of life in seeds include the following phenomena: Removal or burning of a building, disturbance of grassland or forest soils, disturbance of old or ancient mounds, disturbance of the earth's surface by bombing, drainage of old lake or pond beds, and erosion and deposits left by glaciers. Superficially, events arising from these phenomena appear as striking examples of seeds of great longevity. However, such observations are based largely on speculation without considering all the facts.

Turner *(1933)* pointed out the weakness of conclusions based on such observations in the following words: "When the longevity of buried seeds is estimated entirely from the circumstances in which they are found, there is the possibility of error, and the only satisfactory methods of ascertaining the length of time that buried seeds will remain viable are by experiments such as those begun by Beal and Duvel in the United States." Turner *(1933)* described 20 examples of claimed longevity of buried seeds about which the age of the seed can be questioned. We describe five examples, some of which were also described by Turner *(1933).*

Seeds in Soil Disturbed by Digging Graves

In his "Flora of the Summe Battlefield," Hill *(1917)* gave the following account: "In July poppies (*Papaver rhoeas* L.) predominated, and the sheet of colour as far as the eye could see was superb; a blaze of scarlet unbroken by tree or hedgerow. Here and there long stretches of chamomile (*Matricaria chamomilla* L.) broke into the prevailing red and monopolized some acres; and large patches of yellow charlock were also conspicuous, but in the general effect no other plants were noticeable, though a closer inspection revealed the presence of most of the common weeds of cultivation. . . . Charlock (*Sinapis arvensis* L.) not only occurred in broad patches but was also fairly uniformly distributed, though masked by the taller poppies. Numerous small patches were, however, conspicuous and these usually marked the more recently dug graves of men buried where they had fallen. . . . No doubt in the ordinary operations of ploughing and tilling of the ground in years before the war much seed was buried which has been brought to the surface by the shelling of the ground and subsequent weathering. In this connection the presence of charlock on the more recently dug graves, where the chalk now forms the actual surface, is of interest, since it adds further proof of the longevity of this seed when well buried in the soil."

Actually there is no proof that the charlock seeds (*Brassica kaber* (DC.) L. C. Wheeler) (=*Sinapis arvensis*) had been buried either deep or for a long time. The seeds may have been lying in the soil but close to the surface, where conditions were not conducive to germination until they were brought to the surface, where light and oxygen were available and where any accumulated carbon dioxide within the seeds could escape.

Seeds in Soil Under Forests of Known Age

Peter *(1893)* reported studies made of the seed content of soils of a forest that had been planted on meadows and pastures for known periods and kept free from open land plants by shading. In general, as the age of the forests increased, seeds of field plants became more scarce. He found seeds of *Hypericum humifusum, Juncus bufonius,* and *Stellaria media* in deep soil layers of forests 100 years old. In soils of forests 20 to 46 years old he found seeds of a large number of open land plants belonging to such genera as *Anagallis, Chenopodium, Juncus, Plantago, Polygonum, Sinapis, Stachys, Stellaria,* and *Thlaspi.* Peter *(1893)* concluded that seeds of some meadow and swamp plants that may lie in the soil more than 50 years are still capable of germination.

His conclusions appear to be correct, and in lieu of a planned experiment they provide some worthwhile data; however, his data would be more valuable if there were no question about the age of the seeds.

Seeds Frozen in Arctic Tundra

Porsild et al. *(1967)* reported that seeds of arctic lupine, at least 10,000 years old, found in lemming burrows deeply buried in permanently frozen silt of the Pleistocene age, germinated and grew into normal, healthy plants. A mining engineer found the seeds in the permanently frozen burrows associated with the remains of a nest, fecal matter, skulls, and skeletons of a rodent, later identified as the collared lemming. According to Porsild et al. *(1967)*, the mining engineer kept the skeletons in a dry place for 12 years, when the authors learned of the specimens and historical background.

Through carbon-14 dating, comparison with other archeological artifacts, and natural and geological history of the central Yukon area where the seeds were found, they concluded that the seeds had been placed in the burrows by the lemmings and were at least 10,000 years old. About two dozen seeds easily identified as those of the arctic lupine were found, and about half were remarkably well preserved. Some of the best preserved seeds yielded six seedlings when tested for germination and eventually yielded normal, mature plants, indistinguishable from those grown from fresh seeds. The collared lemming no longer occurs in the central Yukon, but the arctic lupine is a common species of tundra and subalpine forests of northern and central Alaska, the Yukon Territory, and the northern Mackenzie District.

A critical review of this account reveals the weaknesses of evidence supporting the claimed age of these seeds.

Canna Seeds in Burial Mound

Researchers in Argentina (Sivori et al., *1968*) reported that three seeds of a *Canna* species (exact species not determined) were obtained from a tomb being excavated in Santa Rose de Tastil. The seeds were inside walnut shells strung in a necklace to form rattles. As determined by carbon-14 dating tests of cameloid bones from overlying strata of the site, the seeds were about 530 years old. This information, supplemented by the fact that no Incaic elements were found in the burial mound, suggests that the city was developed before 1420. Two of the three seeds were tested for germination and subsequent growth. The first seed germinated, but the radicle ceased to grow, even after adding gibberellic acid and indole-3-acetic acid. Another seed was tested under aseptic conditions on a medium containing minerals and growth regulators. It germinated and continued growth after transplantation into quartz gravel with a nutrient solution in a greenhouse. The roots of the plant exhibited irregular geotropic response, even after being transplanted in the greenhouse.

The principal weakness in this account of seed longevity is the accuracy of the carbon-14 dating test and the time relationship of the cameloid bones tested and the *Canna* seeds. (See carbon-14 dating results by Libby *(1951)* and Godwin and Willis *(1964)* for *Nelumbo* seeds in the next section.)

Indian Lotus Seeds Buried in Former Lakebed

Probably the longest claimed longevity for any seed is that recorded by Ohga *(1923)*. He obtained approximately 100-percent germination with seeds of Indian lotus, which he found in a peat bed buried 2 feet deep with loess in the Pulantien River valley in southern Manchuria. The bed was 41 feet above the present water level of the river. Judging from the age of the weeping willows on the bed of this former lake and from the lowering rate of the water level, Ohga *(1923)* concluded that the seeds were probably at least 120 years old. Possibly they were 400 years old or even older judging by the rate of erosion. Although the supply of seeds was limited, Ohga deposited some with the Tohuku Imperial University in Japan, and in 1926 he gave 30 seeds to the British Museum in London.

Later Chaney *(1951)* obtained a few seeds from the Ohga collection and made them available for study in the United States. Records are available (Wester, *1973*) showing that in tests made in Japan, England, and the United States on 156 seeds all germinated except 2. A dry seed from the Sloane collection in the British Museum germinated after 237 years (table 30). Two seeds from the Ohga collection, tested for germination in 1951 by Wester, are illustrated in figure 33.

The age of these *Nelumbo* seeds is perhaps the most controversial subject in the field of seed longevity with a legitimate basis. The discoverer of the original seedbed and seeds estimated the seeds to be at least 120 years old and perhaps as old as 400 years (Ohga, *1923*). Later Chaney *(1951)*, a paleontologist, reevaluated the information on the age of the seedbed and concluded that the seeds might be as much as 50,000 years old.

Arnold and Libby *(1951)* reported that they had determined by carbon-14 dating that the seeds were 1,040 ± 210 years of age. This age was generally accepted until Godwin and Willis *(1964)* conducted carbon-14 dating tests and concluded the correct age to be 100 ± 60 years. Wester *(1973)* reviewed the entire history and arrived at an age of 1,024 years, which agrees with Libby's findings of 1,040 ± 210 years.

There is little question that these seeds are old, but their age has not been established to the satisfaction of biologists. Like seeds of many leguminous species, the seeds of the Indian lotus plant have very hard seedcoats, which resist decay and are impervious to water and gases. Also, the peat in which the seeds were found might have retarded the deterioration of the seedcoat.

Life in Mummified Seeds

The following article on "Wheat 2,000 Years Old" appeared in the Sunday magazine section of a Washington, D.C., daily newspaper, dated January 13, 1957:

"This story begins more than 2,000 years ago, when relatives of a deceased Egyptian nobleman buried some wheat kernels in his tomb, presumably so the departed could grow his own in the land of the dead.

"The nobleman was hardly in any shape for showing as a museum mummy when unearthed recently by a Swiss archeologist, but the wheat seeds appeared to be in good condition. The archeologist, no farmer himself, wondered if they might still sprout under proper cultivation.

"Reviewing his proposal to test the grain under various conditions of soil and clime, the Egyptian government turned thumbs down on experimenting anywhere but in Egypt. But placing agronomy ahead of sovereignty, the scientist whisked a handful of the kernels out of the country.

PN–5414, PN–5415

FIGURE 33.—Seeds of *Nelumbo nucifera* alleged to be over 1,000 years old: *Above,* before absorbing water; *below,* after germination.

"Distributing it to several experimental farmers in Europe, he seems to have scored a major success with the wheat only in Sweden. There, Oscar Johnson of Smaland planted three kernels, carefully nursed them through the season and ended up with several thousand seeds for the next planting. By present count, he has enough to sow 2½ acres. Should he have a successful harvest next year, he expects to reap enough grain to plant a few hundred acres. From then on, the wheat that had laid dormant for ages may be available for others.

"Called "Osiris," the reborn wheat strain has thus far proved to have tremendous hardiness. What other beneficial qualities it may have, Mr. Johnson and other agriculturists the world over are interested in finding out."

Two photographs accompanied the article. The caption to one photograph showing four spikes of branched grain reads as follows: "Buried for centuries in an Egyptian tomb, wheat kernels were unearthed and sown by Oscar John Johnson of Sweden, who produced this initial yield." The other photograph, showing a man who apparently is transplanting seedlings of grain in soil, has the following caption: "Farmer Johnson had only three grains of the 2,000-year-old wheat, so he planted them in a special bed where he could keep a watchful eye on their progress. All three grains prospered in spite of their age. The matured wheat exhibited extreme hardiness, and he reaped enough to plant more than two acres now."

This article was sent to the head of the Swedish Seed Testing Station at Solna, Sweden, requesting that he comment on the article. Following is his reply, dated February 1957:

"This is the same old story that appears in the world press at least once in 10 years. The Egyptians by now know very well that it is a lie. If living wheat kernels are found in ancient tombs, they were put there rather recently by mice or other animals. Cereals deposited there 2,000 years ago have now turned to pure coal and are as dead as they can be. When staying in Egypt 10 years ago I had an opportunity to convince myself personally about this. Since no business can be done with such seeds in Egypt, the kernels are sent to other countries and are sold at fancy prices.

"The special case of Mr. Johnson first circulated in the Swedish press a few years ago and then has been taken up by newspapers and journals in most other countries. About a year ago I had a letter from Canada asking my opinion of it. It is a pity that journalists always copy sensational stories but never disprovals of them! Mr. Johnson's wheat is a variety grown today in Egypt."

Other similar stories of mummy seeds are discussed by Brooke *(1935)*, Luthra *(1936)*, and White *(1946)*.

A form of wheat sold to tourists as "miracle wheat" or "mummy wheat" is a common branched form of *Triticum turgidum*, which grows in southern Europe and northern Africa. It is commonly cultivated as a curiosity (Turner, *1933*). Similarly, fasciated forms of a pea similar to the "mummy pea" were illustrated in a European herbal as early as 1590. One form of the pea was described and named *Pisum umbellatum* (rose or crown pea) in 1771 (White, *1946*).

Turner *(1933)* of Kew Gardens, London, indicated that there was no authenticated evidence that wheat taken from undisturbed Egyptian tombs will germinate. An experiment was made at Kew Gardens with grain from a model granary found in a tomb of the 19th dynasty. Samples were tested under various conditions. The effect of colored glass was tried to induce germination, but after 3 months the grain had turned to dust.

We have confirmed these results in tests made at Beltsville, Md., using mummified grain from the Anatolian excavations at the site of Alishar. Information from the Oriental Institute, University of Chicago, indicates that the material dates back to at least 700 to 1200 B.C. and possibly to about 3000 B.C.

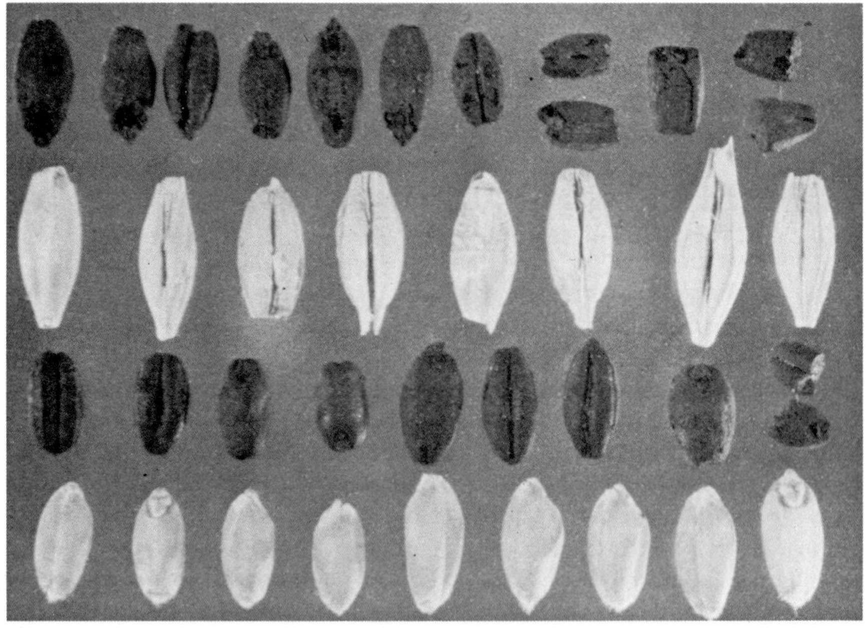

PN–5416

FIGURE 34.—Seeds of 1973 barley and wheat crops compared with carbonized structures ("seeds") of the same species from the Alishar excavations, dating from approximately 700 to 3,000 B.C.: *Top to bottom*, barley from Alishar and 1973 crop; wheat from Alishar and 1973 crop.

The printed word does not seem to dispel the story of life in mummy seeds as such stories appear in the popular press from time to time. For this reason, we are showing some of the seeds from the Alishar excavations along with seeds harvested in 1973 (fig. 34). The so-called mummy seeds have retained the shape of barley and wheat, but the structure is similar to that of charcoal. There is no possibility of these structures producing seedlings. Turner *(1933)* said, "It is popularly asserted that miracle wheat and mummy peas originate from Egyptian tombs and that such seeds germinate when sown, but in every instance the statements prove to be without foundation."

GLOSSARY

Absolute humidity.—Amount of water vapor actually in the air, expressed either in its expansive force or in its weight per given volume, as grains per cubic foot.

Absorption.—Imbibing of water by living cells or tissues in a seed.

Accelerated aging test.—Intentionally subjecting seeds to adverse storage conditions for a short period to estimate their possible lifespan under favorable storage conditions.

Achene.—Small, dry, one-seeded fruit with thin distinct wall, which does not split open.

Actinomycetes.—Bacterialike micro-organisms with ribbonlike form.

Activated alumina.—Highly porous, granular form of aluminum oxide, with high adsorptive capacity for moisture.

Adsorption.—Taking up a gas, vapor, or dissolved material on the surface of a solid; finely divided materials, as active carbon and silica gel, can adsorb relatively large quantities of other materials.

Aeration.—Movement of outside air through stored seeds to prevent heating and to facilitate drying.

Aging test.—Any test to determine the deterioration of seeds with their increased age, usually a germination or vigor test.

Air-dry.—Sufficiently dry so that no further moisture is given off on exposure to air; i.e., in moisture equilibrium with the surrounding air.

Air-oven method.—Drying seeds in a forced-air oven to determine their moisture content.

Alpha amylase.—An enzyme occurring widely in seeds, leaves, and other plant parts; it accelerates starch hydrolysis.

Ambient.—That which encompasses or surrounds on all sides a designated object, state of matter, system, etc.; as used here, ambient refers to temperature or relative humidity of the atmosphere in the natural state without modification by man.

Anabolic process.—Process by which matter is changed into tissues of living plants or seeds.

Axillary bud.—Bud developed in the angle between leaf and stem.

Backcrossing.—Crossing a hybrid with one of the original parents.

Bacteria.—Simple microscopic vegetable organisms usually single-celled and without chlorophyll, multiplying by fission (splitting apart) and spore formation.

"Baldheads."—Bean seedlings with the primary leaves missing and the terminal bud present or absent.

Batch drying.—Drying seeds in relatively small quantities while stationary as opposed to drying in a continuous moving line.

Biochemical change.—Change in the chemistry of a plant, plant part, or seed.

Bracts.—More or less modified leaves subtending a flower or belonging to an inflorescence.

Bruchids.—Family of small vegetarian beetles; the larvae live mainly in the seeds of leguminous plants.

Bunt spores.—Reproductive units of certain fungi that attack cereal plants.

Caryopsis.—One-seeded fruit with pericarp and seedcoat fused into one covering, as in corn and other grains and grasses.

Catabolic process.—Pertaining to or characterized by the release of energy.

Chaff.—Glumes or husks of grains and grasses separated from the seed by threshing, winnowing, etc.

Chalcids.—Minute wasplike insects belonging to the order Hymenoptera, a few species of which invade seeds.

Check samples.—Samples of original seed lots used as a basis for establishing the effects of any special treatment on the response of seeds of the same lots. For example, seeds held at room temperature constitute the check sample for seeds of the same lot stored at a high temperature and high relative humidity and at a low temperature and low relative humidity.

Chromosome aberrations.—Deviations from normal chromosome structure.

Chromosome 10.—Geneticists have assigned numbers to various chromosomes; 10 identifies a specific chromosome.

Closed system.—Completely enclosed, sealed system; device, container, or combination of devices designed to perform a specific function within a sealed environment, as the refrigeration system for a domestic refrigerator.

"Cold Test."—Special germination test; for corn, seeds are placed in a very moist sand/soil/peat moss mixture, then held at 10° C for 7 days before being placed in the regular germination temperature.

Coleoptile.—Sheathlike leaf of grasses and other monocotyledons; it protects the delicate growing point as it emerges from the soil.

Coleorhiza.—Sheath surrounding the radicle in some plants, through which the root bursts.

Combine thresher.—Machine that cuts, threshes, and cleans grain while moving over the field.

Convection.—A type of air current.

Cotyledons.—Seed leaves of the embryo, usually thickened for storage of reserve food; they may serve as true foliage leaves.

Cultivar.—Cultural or cultivated variety.

Culture medium.—Laboratory substance on which micro-organisms and excised embryos may be grown.

Cuticle.—Very thin detachable skin covering a plant, especially the leaves.

Cytological change.—Change in the intrinsic character or function of a cell.

Decorticated seed.—Seed with its outer covering removed.

Dehumidification.—Process by which the water is removed from a substance; moisture vapor is removed from air.

Dehumidifier.—Equipment used to remove humidity (moisture) from air.

Desiccant.—Drying agent.

Desiccator.—Usually a glass jar fitted with an airtight cover and containing at the bottom some desiccating agent, such as calcium chloride.

Desorption.—Removal of a substance from an absorbed state.

Dielectric properties.—Properties that make a material a nonconductor of electricity.

Dough stage.—Stage in development of cereal grains when the interior of the kernel (seed) is of doughlike consistency.

Dry-bulb thermometer.—Ordinary thermometer with its mercury-containing bulb kept dry. (See psychrometer.)

Dry heat.—Heat applied by dry air as opposed to moist heat, which is applied by moist air.

Dry weight.—Moisture-free weight.

Dynamic method.—Procedure for determining equilibrium moisture content by bubbling air through absorption towers containing an acid or saturated salt solution, which controls the humidity around seeds or other material.

Economic plants.—Plants of practical utility; commercial plants.

Ecosystem.—System involving all ecological or environmental factors, such as climate, soil, physiography, and associated organisms.

Electric hygrometer.—Electric instrument for measuring the degree of moisture in the atmosphere.

Electric moisture meter.—Electrical instrument that measures the moisture content of seeds and other materials.

Embryo.—Rudimentary plant within the seed.

Embryo axis.—Hypocotyl-root axis bearing at one end the root meristem and at the other the cotyledon or cotyledons and the shoot meristem.

Embryo transplant technique.—Procedure by which the embryo of one seed is transplanted to the endosperm of another seed.

Endogeocarpic flora.—Certain microscopic plants that live or are found in the soil and can invade plants.

Endosperm.—Tissue of seeds, developing from fertilization of the polar nuclei of the ovule by the second male nucleus; it nourishes the embryo.

Enzyme.—Catalyst produced in living matter; a specialized protein capable of aiding in bringing about chemical changes; promotes a reaction without itself being changed or destroyed.

Ergot.—Hard resting body produced by the fungus *Claviceps purpurea*; also the name of a disease of rye and other grasses.

Field emergence.—Emergence of seedlings from seeds planted in the field; usually refers to percentage of seeds planted in the field that germinates and produces seedlings strong enough to emerge through the soil.

Field fungi.—Molds that live outside on growing crops.

Food transport system (in embryo).—System by which food is moved from storage tissues of the seed to growing tips of the root and shoot.

Forage.—Vegetable food of any kind for animals, especially that consumed by domestic animals as pasture or browse.

Freeze-drying.—Process for drying seeds and other materials; it uses a freezing temperature combined with moving dry air, which sublimes frozen water to water vapor.

Fruit(s).—Ripened (matured) ovary of a plant, together with any intimately attached parts that developed with it from the flower. Fruits of buckwheat, cereals, grasses, sunflower, and many others are commonly called seeds.

Fumigant.—Chemical applied as a gas to destroy molds, insects, rodents, etc.

Fungi.—Lower plant forms devoid of chlorophyll and comprising molds, mildews, rusts, smuts, mushrooms, toadstools, puffballs, and allied forms that reproduce by spores.

Fungicide.—Chemical used to kill fungi.

Gall(s).—Swollen plant part caused by presence or action of an insect or nematode.

Gaseous water.—Water vapor.

Gas storage.—Storage in an atmosphere composed of a gas other than air.

Genera.—Plural of genus; groups of structurally or phylogenetically related species; categories of classification ranking between family and species.

Genetic stock.—Strain of seeds used in studies pertaining to the genesis of a particular plant; its mode of production and development; strain used for genetic studies.

Genotype.—Hereditary makeup of an individual plant, which, with the environment, controls the individual's characteristics, such as type of flower or shape of leaf.

Germinate.—Initiate growth or development, especially when referring to a spore or seed.

Germ plasm.—Living substance of the cell nucleus that determines the hereditary properties of organisms and that transmits these properties from makeup of organisms. For example, geneticists and plant breeders often refer to the seeds and plants used in their research and breeding as their "collection of germ plasm."

Glumes.—Chafflike bracts; specifically, two empty chaffy bracts at the base of spikelets in grasses.

Grain.—Seed or seedlike fruit of any cereal grass, such as millet, oats, rice, and wheat.

Grain head.—Attachment for a combine designed specifically for harvesting grain.

Hair hygrometer.—Instrument in which expansion and contraction of a hair is used to measure changes in the amount of moisture in the air.

Hard seed.—Seed with seedcoat impervious to water or oxygen required for germination, or to carbon dioxide that may accumulate within the seed and retard or prevent germination; sometimes overcome by scratching or scarifying the seedcoat or removal by brief immersion in concentrated sulfuric acid and thorough washing.

Hectare.—1 ha is 2.4710 acres.

Herbarium specimen.—Dried and pressed specimens of plants mounted or otherwise prepared for permanent preservation and systematic arrangement in a room, building, or institution (herbarium) where such collections are kept.

Hermetically sealed.—Sealed so that no gas or moisture can enter or escape.

Hermetic storage.—Storage in an airtight container.

Hilum.—Scar or point of attachment of the seed.

"Hot spots."—Areas in a mass of stored seeds or grain where temperature is as high as 66° C or higher.

Hybrid.—Offspring of the union of a male of one race, variety, species, genus, etc., with a female of another; a crossbred plant.

Hydrophobic.—Lacking a strong affinity for water.

Hygroscopic.—Readily absorbing and retaining moisture, as caustic potash, or becoming coated with water, as glass.

Hygroscopic equilibrium measurement.—Measuring the amount of water in or on a substance when there is no movement of water between the substance and the surrounding air.

Hygroscopicity.—Denotes sensitivity to moisture.

Hygroscopic moisture absorption.—Absorption of moisture from the surrounding air.

Hygrothermograph.—Instrument that records both humidity and temperature on the same chart.

Hysteresis effect.—Phenomenon when at a given relative humidity seeds or grain may reach two different equilibrium moisture contents—one by increasing the relative humidity from a low level and another by decreasing the relative humidity from a high level.

Illumination.—Supplying light.

Immature seed.—Seed not fully developed.

Inbred.—Successively self-fertilized; also, a plant or progeny resulting from successive self-fertilization.

Inert.—Lifeless.

Interstitial air spaces.—Air spaces between the cells in a plant tissue or inside a seed.

Layer drying.—Drying seeds in thin layers as opposed to a large mass of seeds.

Leguminous plants or legumes.—Members of the pea or legume plant family (Leguminosae).

Lemma.—Small greenish bract that is part of the floret in grasses.

Lifespan.—Length of time a particular lot or kind of seeds will survive under favorable storage conditions.

Lipids.—Group of substances comprising the fats and other esters that possess analogous properties.

Lipolytic degradation.—Decomposition of fat.

Longevity.—Long duration of life, length of life, or lifespan.

Luteus genes.—Genes that carry a yellow plant color factor; they are lethal as they prevent normal chlorophyll development.

Macromolecular water layer.—Water layer surrounding a macromolecule.

Macromolecule.—Large molecule; grouping together of simple molecules capable of independent existence as in cellulose.

Malt agar.—Culture medium used for laboratory cultivation of certain fungi.

Malvaceous plant.—Plant belonging to the mallow family, as cotton and okra.

Manifold.—Chamber from which air is distributed to more than one duct.

Mature stage.—Completely developed; full grown.

Membrane.—Any thin, soft, pliable sheet or layer of animal or vegetable origin; a limiting protoplasmic surface.

Mercurial.—Chemical compound containing mercury.

Metabolism.—Chemical changes occurring in an organic body in living and growing processes.

Metabolites.—Products of metabolism.

Microflora.—Microscopic plants as a whole that inhabit a given area, such as a seed.

Micropyle.—Pore or opening through which the pollen tube enters the embryo sac during fertilization and through which the radicle frequently emerges when the seed germinates.

Milk stage.—Stage in the development of the kernel of a grain when its contents are a milklike liquid.

Moisture-barrier material.—Any material through which moisture either cannot move or moves very slowly.

Moisture content equilibrium.—State when the moisture content in the air and in the seed remains constant; no movement of moisture either into or out of the seed.

Moisture equilibrium.—Relative humidity of the air is in equilibrium with the moisture content of the seed.

Moisture gradient.—Difference in moisture concentration inside and outside a seed and between the seed surface and the air surrounding the seed.

Moisture translocation.—Movement of moisture from one part of a seed to another.

Multiwall bag.—Bag with multiple walls, usually two or more, composed of paper, foil, plastic film, or cloth, alone or in various combinations.

Mutagenic changes.—Change in the genetics of a seed; change in form or qualities of chromosomes.

Mycoflora.—Fungal component of a microflora.

Nematode.—Microscopic wormlike animal.

Nomograph.—Alinement chart; graph that enables one, by the aid of a straight edge, to read the value of a dependent variable when the value of the independent variable is given.

Nutlet.—Any small nutlike fruit or seed, as in the mint and vervain plant families.

Oilseed.—Any seed from which oil is expressed; a seed containing a large percentage of oil in its storage tissues.

"Once-over harvesting."—Harvesting the entire crop at one time, as opposed to going over a field several times to complete the harvest.

Open storage.—Storage with free access to normal atmospheric conditions.

Ordinary storage.—Storage as usually practiced with no attempt to control either temperature or relative humidity of the storage area.

Osmophilic.—Preferring to grow with very little water.

Overwinter storage.—Storage through the winter from harvest in the fall to planting time in the spring.

Ovules.—Bodies within the ovary of a flower that become seeds after fertilization and development.

Oxidation.—Act or process of being oxidized.

Palea.—Upper bract that with the lemma encloses the flower in grasses. (See glumes.)

Pallet box.—Box in which seeds are placed for handling during processing and that is moved by forklift. (See totebox.)

Parasitic fungi.—Molds that exist at the expense of other organisms.

Percent germination.—Percentage of seeds that produces normal seedlings when tested in a laboratory according to established procedures.

Pericarp.—Seed covering derived from the ovary wall; it may be thin and intimately attached to the seedcoat as in a kernel of corn, fleshy as in berries, or hard and dry as in pods and capsules.

Photoperiod.—Length of the light period on a given day.

Photoperiodism.—Response to the timing of light and darkness.

Photosynthetic organs.—Organs containing chlorophyll and capable of photosynthesis.

Physiological dormancy.—Dormancy caused by disorganization of functions or of metabolism as brought about by extremes of temperature, oversupply of water, etc.

Placenta.—Any sporangia-bearing surface; in seed plants that part of the carpel bearing ovules.

Planting stock.—Seeds selected and preserved especially for planting, as opposed to seeds held for human or animal food.

Plumule.—Major young bud of the embryo within a seed or seedling from which the aerial part of the plant will develop.

Premilk stage.—Stage in the development of a seed that precedes the milk stage.

Primary root.—First root of a plant that develops from the radicle.

Propagation.—Perpetuating and increasing plants vegetatively.

Protoplasm.—Essential, complex, living substance of cells on which all the vital functions of nutrition, secretion, growth, and reproduction depend.

Provenance.—Origin, source, place where found.

Psychrometer.—Hygrometer or instrument for measuring the water vapor in the atmosphere, consisting essentially of two similar thermometers, the bulb of one kept dry, the other wet. The wet bulb is cooled as a result of evaporation and consequently shows a temperature lower than that of the dry-bulb thermometer. Since evaporation is less on a moist day, the difference between the thermometer readings is greater when the air is dry. This difference constitutes a measure of the dryness of the surrounding air.

Psychrometric chart.—Graphical representation of psychrometric data, showing the relationship between dry-bulb temperature, dewpoint, wet-bulb temperature, and relative humidity.

Radicle.—Rudimentary root or lower end of the hypocotyl of the embryo; it forms the primary root of the young plant.

Relative humidity.—Ratio of the quantity of water vapor actually in the air to the greatest amount possible at a given temperature.

Respiration.—Oxidation-reduction process occurring in all living cells whereby chemical action occurs, producing compounds and releasing energy that are partly used in various life processes.

Respiration quotient (RQ).—Ratio of the volume of carbon dioxide (CO_2) given off in respiration to that of the oxygen (O_2) consumed—$RQ = CO_2/O_2$.

Root axis.—Principal line along which a root may grow; hypothetical central line of a root.

Sample.—Part of a seed lot presented for inspection or shown as evidence of the quality of the whole lot.

Saprophytic fungi.—Molds that live off nonliving material.

Scarified seed.—Seed that has had its coat scratched to permit ready entry of water and exchange of gases.

Schematic flow.—Movement of seeds according to a previously designed scheme or pattern.

Sclerotia.—Hard, resting bodies produced by certain fungi.

Scutellum.—Specially differentiated cotyledon in grass seeds; shield-shaped organ through which the embryo absorbs food from the endosperm.

Sealed storage.—Storage in a sealed container; usually refers to hermetic storage.

Secondary root.—Lateral branch of the primary root.

Seed(s).—Mature ovule consisting of an embryonic plant together with stored food, all surrounded by a protective coat.

Seedcoat.—Outermost tissues or “skin” of a seed; sometimes this coat is extremely hard and waterproof, preventing entrance of water to initiate germination unless it is broken, scratched, or eroded.

Seedling.—Young plant grown from a seed.

Seed lot.—Specific quantity of seed usually assigned an identifying number or symbol; examples are the seed from a single field or a day’s harvest.

Seed physiology.—Branch of biology dealing with the organic processes and phenomena collectively of seeds and their parts.

Seed stock.—Quantity of seed reserved for planting for the production of a new seed crop.

Shoot axis.—Principal line along which a shoot may grow; hypothetical central line of a shoot.

Sigmoid.—Curved in two directions, like the letter "S."

Silica gel.—Regenerable adsorbent consisting of amorphous silica.

Smut ball.—Reproductive body produced by smut fungi.

Species.—Group of closely related organisms; for example, *Medicago sativa* is the botanical name for alfalfa. *Medicago* is the genus and *sativa* is the species. Several species belong to the genus *Medicago*; for example, *M. lupulina* (black medic) and *M. orbicularis* (button-clover).

Spine(s).—Any stiff sharp process distinguished from a thorn by the absence of vascular tissue. Spines are frequently found on leaf margins as in thistles.

Stand of plants.—Relative number or distribution of plants growing over a given area, especially soon after germination.

Static efficiency.—Efficiency of static pressure in moving air through a layer of seeds.

Static method.—Method by which a large fan builds up air pressure against a mass of seeds in a dryer in order to move more air through the seeds so as to speed up the drying process.

Static pressure.—Pressure acting by mere weight (force) without motion; pressure exerted by mere mass at rest.

Storage fungi.—Fungi that live off stored seeds and grains.

Stress test.—Test in which seeds are intentionally subjected to the action of adverse external forces, usually unfavorable germination conditions.

Strophiole.—Swollen appendage at the hilum of certain seeds, as that of spurge.

Substrate.—Substratum, a substance acted upon.

Substratum.—Seedbed in a germination test; soil is a substratum for germinating seeds.

Temperature gradient.—Rate of temperature change with increase in height; lapse rate or change with distance from heat or cooling source.

Terminal bud.—Bud growing at the end of a branch or stem.

Testa.—Hard external coating or integument of a seed; seedcoat.

Thermocouple.—Thermoelectric couple used to measure temperature differences.

Threshed.—Beat out from the head, pod, or capsule; separated from the plant.

Totebox.—Open or closed metal or wood box, which is placed on a pallet and is moved by a forklift. (See pallet box.)

Treated seed.—Seed to which a poisonous chemical compound has been applied to kill pests.

Umbel(s).—Inflorescence in which the peduncles or pedicels of a cluster spring from the same point, as in carrot, coriander, or parsnip.

Unthreshed.—Not beat out; not separated from the pod, head, or capsule.

Vacuum storage.—Storage in a sealed container from which all or nearly all air was evacuated before sealing; storage in a (partial) vacuum.

Vascular.—Furnished with vessels or ducts.

Viability.—Quality as state of being alive; ability to live, grow, and develop, as the viability of certain grains under dry conditions.

Viability test.—Test to determine whether a seed or the percentage of seeds in a lot is alive or dead.

Viable.—Capable of growing or developing, as viable seeds.

Vigor.—Condition of active good health and natural robustness; when seeds are planted, vigor permits germination to proceed rapidly and to completion under a wide range of conditions.

Virus.—Extremely small, filterable body, which multiplies only in living cells.

Vitality.—Germinating power.

Water-curtain germinator.—Germinator (germination chamber) that is heated, cooled, aerated, and humidified by means of a curtain of water falling down the back of the chamber; water is recirculated.

Weevils.—Snout beetles of the suborder Rhynchophora with snout-bearing jaws at the tip that damage stored seeds and grain.

Wet-bulb thermometer.—Ordinary thermometer with mercury-containing bulb that is kept wet. (See psychrometer.)

Wet weight.—Weight before drying.

Windrow.—Row of hay raked up to dry before being stacked or baled; also any similar row for drying, as of grain or peanuts.

Xerophyte.—Plant structurally adapted for growth with a limited water supply.

Yellow cuprocide.—Copper-containing fungicide used to protect seeds from soil fungi.

Yield.—Aggregate of products resulting from growth or cultivation; quantity produced, such as 100 bushels of corn per acre.

Zygote.—Fertilized egg, potentially capable of developing into a seed.

VERNACULAR AND BOTANICAL-ZOOLOGICAL NAME EQUIVALENTS[11]

Plants

Achillea, the pearl—*Achillea ptarmica* L.
Ageratum—*Ageratum mexicanum* Sims
Alfalfa—*Medicago sativa* L.
Alyceclover—*Alysicarpus vaginalis* (L.) DC.
Alyssum—*Alyssum maritimum* (L.) Lam.
Amaranth, redroot—*Amaranthus retroflexus* L.*
Amaranthus—*Amaranthus* spp.
Anemone—*Anemone* spp.
Angel-trumpet—*Datura arborea* L.
Anise—*Pimpinella anisum* L.
Apple—*Malus sylvestris* Mill.
Arabis—*Arabis alpina* L.
Armeria—*Armeria* spp.
Artichoke—*Cynara scolymus* L.
Asparagus—*Asparagus officinalis* L.
Asparagus, fern—*Asparagus plumosus* Baker and *A. sprengeri* Regel
Asparagusbean, sitao—*Vigna unguiculata* subsp. *sesquipedalis* (L.) Verdc.
Aster—*Aster* spp.
Aster, China—*Callistephus chinensis* (L.) Nees
Babysbreath—*Gypsophila* spp.
Bachelor's button, cornflower—*Centaurea cyanus* L.
Bahiagrass—*Paspalum notatum* Fluegge
Balloonflower—*Platycodon grandiflorum* DC.
Balm—*Melissa officinalis* L.
Balsam—*Impatiens* spp.
Barley—*Hordeum vulgare* L.
Barnyardgrass, junglerice—*Echinochloa* spp.*
Basil, sweet—*Ocimum basilicum* L.
Basketflower—*Centaurea americana* Nutt. and *Hymenocallis calathina* Nichols
Bean:
    Field—*Phaseolus vulgaris* L.
    Garden (snap)—*Phaseolus vulgaris* L.
    Kidney—*Phaseolus vulgaris* L.
    Lima—*Phaseolus vulgaris* L.
    Mung—*Vigna radiata* (L.) Wilczek (syn. *Phaseolus aureus* Roxb.)

---

[11] Since vernacular names are not commonly used for storage fungi, these organisms are not listed. * = only scientific name is given in text.

Bean—Continued

Navy—*Phaseolus vulgaris* L.

Scotch—*Vicia faba* L.

Tapilan—*Vigna umbellata* (Thunb.) Ohwi and Ohashi (syn. *Phaseolus calcaratus* Roxb.)

Beet:

Field—*Beta vulgaris* L.

Garden—*Beta vulgaris* L.

Sugar—*Beta vulgaris* L.

Beggarweed—*Desmodium tortuosum* (Sw.) DC.

Begonia—*Begonia* spp.

Bellflower:

Bluebell—*Campanula* spp.*

Peach—*Campanula persicifolia* L.

Bentgrass:

Colonial (including Astoria and Highland)—*Agrostis tenuis* Sibth.

Creeping—*Agrostis palustris* Huds.

Velvet—*Agrostis canina* L.

Bermudagrass—*Cynodon dactylon* (L.) Pers.

Betony—*Stachys* spp.*

Bindweed—*Convolvulus* spp.*

Bluegrass:

Bulbous—*Poa bulbosa* L.

Canada—*Poa compressa* L.

Kentucky—*Poa pratensis* L.

Nevada—*Poa nevadensis* Vasey ex Scribn.

Rough—*Poa trivialis* L.

Texas—*Poa arachnifera* Torr.

Wood—*Poa nemoralis* L.

Bluestem:

Big—*Andropogon gerardii* Vitm.

Little—*Andropogon scoparius* Michx.

Sand—*Andropogon hallii* Hack.

Borage—*Borago officinalis* L.

Broadbean or horsebean—*Vicia faba* L.

Broccoli—*Brassica oleracea* var. *botrytis* L.

Bromegrass:

Mountain—*Bromus marginatus* Nees ex Steud.

Smooth—*Bromus inermis* Leyss.

Browallia—*Browallia* spp.

Brussels sprouts—*Brassica oleracea* var. *gemmifera* DC.

Buckwheat—*Fagopyrum esculentum* Moench

Buffalograss—*Buchloe dactyloides* (Nutt.) Engelm.

Buffelgrass—*Cenchrus ciliaris* L. (syn. *Pennisetum ciliare* (L.) Link)

Bugloss—*Anchusa capensis* Thunb.
Burclover:
California—*Medicago polymorpha* L.
Spotted—*Medicago arabica* (L.) Huds.
Burdock—*Arctium lappa* L. and *A. minus* (Hill) Bernh.
Bushclover—*Lespedeza violacea* (L.) Pers.
Butterflyflower—*Schizanthus wisetonensis* Hort.
Buttonclover—*Medicago orbicularis* (L.) Bartal.
Cabbage—*Brassica oleracea* var. *capitata* L.
Cabbage, Chinese—*Brassica pekinensis* (Lour.) Rupr.
Calceolaria—*Calceolaria* spp.
Calendula—*Calendula officinalis* L.
Calliopsis, dwarf and tall—*Calliopsis* spp.
Canarygrass—*Phalaris canariensis* L.
Canarygrass, reed—*Phalaris arundinacea* L.
Candlenut—*Aleurites moluccana* (L.) Willd.*
Candytuft:
Annual—*Iberis umbellata* L.
Perennial—*Iberis gibraltarica* L. and *I. sempervirens* L.
Canna—*Canna* spp.
Cantaloup—*Cucumis melo* L.
Canterbury-bells—*Campanula medium* L.
Caraway—*Carum carvi* L.
Cardoon—*Cynara cardunculus* L.
Carnation—*Dianthus caryophyllus* L.
Carpetgrass—*Axonopus affinis* Chase
Carrot—*Daucus carota* L.
Castorbean, Cambodia—*Ricinus cambodgensis* Benary*
Cathedral bells—*Cobaea scandens* Cav.
Catnip, nepeta—*Nepeta* spp.
Cauliflower—*Brassica oleracea* var. *botrytis* L.
Celeriac—*Apium graveolens* var. *rapaceum* (Mill.) Gaud.
Celery—*Apium graveolens* var. *dulce* (Mill.) Pers.
Centaurea:
Royal—*Centaurea imperialis* Hort.
Velvet—*Centaurea gymnocarpa* Moris and Denot
Chamomile, German (sweet false)—*Matricaria chamomilla* L.
Chard, Swiss—*Beta vulgaris* var. *cicla* L.
Charlock—*Sinapis arvensis* L. = (*Brassica kaber* (DC) L. C. (Wheeler))
Chervil—*Anthriscus cerefolium* (L.) Hoffm.
Chickweed, common—*Stellaria media* (L.) Vill.*
Chicory—*Cichorium intybus* L.
Chives—*Allium schoenoprasum* L.
Chrysanthemum, annual—*Chrysanthemum* spp.

Cinchona—*Cinchona ledgeriana* Moens ex Trim.*
Cineraria, common—*Senecio cruentus* (L'Her.) DC.
Cleome, spiderflower—*Cleome gigantea* Hort.
Clover:
Alsike—*Trifolium hybridum* L.
Berseem—*Trifolium alexandrinum* L.
Crimson—*Trifolium incarnatum* L.
Ladino—*Trifolium repens* L.
Lappa—*Trifolium lappaceum* L.
Large hop—*Trifolium procumbens* L.
Persian—*Trifolium resupinatum* L.
Red—*Trifolium pratense* L.
Rose—*Trifolium hirtum* All.
Strawberry—*Trifolium fragiferum* L.
Striate—*Trifolium striatum* L.*
Subterraneum—*Trifolium subterraneum* L.
Suckling (small hop)—*Trifolium dubium* Sibth.
White—*Trifolium repens* L.
Cockscomb—*Celosia* spp.
Cocksfoot (orchardgrass)—*Dactylis glomerata* L.
Cockvine, thunbergia—*Thunbergia alata* Bojer
Cocoa—*Theobroma cacao* L.
Coleus, common—*Coleus blumei* Benth
Collards—*Brassica oleracea* var. *acephala* DC.
Coltsfoot—*Tussilago farfara* L.*
Columbine—*Aquilegia* spp.
Coneflower—*Rudbeckia* spp.
Coralbells—*Heuchera sanguinea* Engelm.
Coreopsis, perennial—*Coreopsis lanceolata* L.
Coriander—*Coriandrum sativum* L.
Corn (dent, field, flint, popcorn, sweet, white, yellow)—*Zea mays* L.
Cosmos—*Cosmos* spp.
Cotton—*Gossypium* spp.
Cowpea—*Vigna unguiculata* (L.) Walp. subsp. *unguiculata*
Cranesbill—*Geranium* spp.
Cress:
Garden—*Lepidium sativum* L.
Water—*Nasturtium officinale* R. Br.
Crested dogtail—*Cynosurus cristatus* L.
Crotalaria—*Crotalaria intermedia* Kotschy, *C. juncea* L., *C. lanceolata* E. Mey., *C. spectabilis* Roth, and *C. striata* DC.
Crownvetch—*Coronilla varia* L.
Cucumber—*Cucumis sativus* L.
Cyclamen—*Cyclamen europeum* L.

Cypressvine—*Ipomoea quamoclit* L.
Dahlia—*Dahlia* spp.
Daisy:
African—*Dimorphotheca aurantiaca* DC.
African lilac—*Arctotis grandis* Thunb.
English—*Bellis perennis* L.
Painted—*Pyrethrum* spp.
Shasta—*Chrysanthemum leucanthemum* L.
Swan river—*Brachycome* spp.
Dallisgrass—*Paspalum dilatatum* Poir.
Dames rocket, sweet rocket—*Hesperis matronalis* L.
Dandelion—*Taraxacum officinale* Weber
Delphinium, annual, perennial—*Delphinium* spp.
Dichondra—*Dichondra repens* Forst.
Dill—*Anethum graveolens* L.
Dock, curly—*Rumex crispus* L.*
Dodder—*Cuscuta* spp.*
Dropseed, sand—*Sporobolus cryptandrus* (Torr.) A. Gray
Dusty-miller—*Centaurea candidissima* Lam.
Eggplant—*Solanum melongena* L.
Endive—*Cichorium endivia* L.
Evening primrose—*Oenothera biennis* L. and *O. lamarckiana (O. grandiflora* Ait.)*
Fennel—*Foeniculum vulgare* Mill.
Fescue:
Chewings—*Festuca rubra* subsp. *commutata* Gaud.
Creeping red—*Festuca rubra* L.
Hair—*Festuca tenuifolia* Sibth.
Meadow—*Festuca pratensis* Huds. (syn. *F. elatior* L.)
Sheep—*Festuca ovina* L.
Tall—*Festuca arundinacea* Schreb.
Firebush, Mexican—*Kochia* spp.
Flax:
Common—*Linum usitatissimum* L.
Flowering—*Linum grandiflorum* Desf.
Perennial—*Linum perenne* L.
Forget-me-not—*Myosotis* spp.*
Foxglove—*Digitalis* spp.
Foxtail—*Setaria* spp.*
Gaillardia—*Gaillardia* spp.
Geranium—*Geranium* spp.
Geum—*Geum* spp.
Gilia—*Gilia* spp.
Globe amaranth—*Gomphrena globosa* L.

Gloxinia, common—*Sinningia speciosa* (Lodd.) Hiern
Godetia—*Godetia amoena* Lilija and *G. grandiflora* Lindl.
Goodia clover—*Goodia latifolia* Salisb.*
Goosegrass—*Eleusine indica* (L.) Gaertn.*
Gourds—*Cucurbita* spp. and *Lagenaria* spp.
Grama:
    Blue—*Bouteloua gracilis* (H.B.K.) Lag.
    Side oats—*Bouteloua curtipendula* (Michx.) Torr.
Guar—*Cyamopsis tetragonoloba* (L.) Taub.
Guineagrass—*Panicum maximum* Jacq.*
Hardinggrass—*Phalaris tuberosa* var. *stenoptera* (Hack.) Hitchc.
Hawksbeard—*Crepis* spp.*
Heliopsis—*Heliopsis* spp.
Heliotrope—*Heliotropium* spp.
Hemp—*Cannabis sativa* L.
Hibiscus—*Hibiscus* spp.
Hollyhock—*Althaea rosea* (L.) Cav.
Hung tau—*Vigna radiata* (L.) Wilczek*
Hyacinth-bean—*Dolichos lablab* L.
Hyssop—*Hyssopus officinalis* L.
Indiangrass, yellow—*Sorghastrum nutans* (L.) Nash
Indigo—*Indigofera cytisoides* L.*
Indigo, hairy—*Indigofera hirsuta* L.
Iris, Japanese—*Iris kaempferi* Sieb.
Japanese lawngrass—*Zoysia japonica* Steud.
Jerusalem or Maltese cross—*Lychnis chalcedonica* L.
Jimsonweed—*Datura* spp.*
Jobs-tears—*Coix lacryma-jobi* L.
Johnsongrass—*Sorghum halepense* (L.) Pers.
Jute—*Corchorus capsularis* L. and *C. olitorius* L.
Kale—*Brassica oleracea* var. *acephala* DC.
Kenaf—*Hibiscus cannabinus* L.
Kidneyvetch—*Anthyllis vulneraria* L.*
Kohlrabi (knolkol)—*Brassica oleracea* var. *gongylodes* L.
Kudzu—*Pueraria lobata* (Willd.) Ohwi (syn. *P. thunbergiana* (Sieb. and Zucc.) Benth.)
Lambsquarters, pigweed—*Chenopodium* spp.*
Lantana—*Lantana camara* L.
Larkspur:
    Annual—*Delphinium ajacis* L.
    Hybrids—*Delphinium hybridum* Steph.
Leadtree—*Leucaena leucocephala* (Lam.) de Wit*
Leek—*Allium porrum* L.
Lentil—*Lens culinaris* Medic.

Lespedeza:

Kobe—*Lespedeza striata* (Thunb.) Hook. and Arn.

Korean—*Lespedeza stipulacea* Maxim.

Sericea or Chinese—*Lespedeza cuneata* (Dumont) D. Don

Siberian—*Lespedeza hedysaroides* (Pallas) Ricker

Striate—*Lespedeza striata* (Thunb.) Hook. and Arn.

Lettuce—*Lactuca sativa* L.

Lily—*Lillium* spp.

Linaria—*Linaria* spp.

Lobelia—*Lobelia erinus* L. and *L. cardinalis* L.

Locustbean (carob)—*Ceratonia siliqua* L.

Lotus, Indian—*Nelumbo nucifera* Gaertn.

Lovegrass, weeping—*Eragrostis curvula* (Schrad.) Nees

Lunaria, honesty—*Lunaria annua* L.

Lupine:

Annual types—*Lupinus* spp.

Arctic—*Lupinus arcticus* S. Wats.

Blue—*Lupinus angustifolius* L.

Russell hybrids—*Lupinus polyphyllus* Lindl.

White—*Lupinus albus* L.

Yellow—*Lupinus luteus* L.

Maize (corn)—*Zea mays* L.

Mallow—*Malva* spp.

Manilagrass—*Zoysia matrella* (L.) Merr.

Marigold:

African—*Tagetes erecta* L.

French—*Tagetes patula* L.

Marjoram—*Origanum majorana* L.

Marvel of Peru, four-o'clock—*Mirabilis jalapa* L.

Matricaria—*Matricaria* spp.

Meadow foxtail—*Alopecurus pratensis* L.

Medic—*Medicago* spp.

Medic, black—*Medicago lupulina* L.

Mignonette—*Reseda odorata* L.

Milkvetch—*Astragalus massiliensis* Lam. *(A. tragacantha)* and *A. utriger* Pallas*

Millet:

Foxtail (common, German, golden, Hungarian, Siberian)—*Setaria italica* (L.) Beauv.

Japanese—*Echinochloa crusgalli* var. *frumentacea* (Link) W. F. Wight

Pearl—*Pennisetum americanum* (L.) Leeke

Proso—*Panicum miliaceum* L.

Mimosa—*Mimosa glomerata* Forsk.*

Morningglory—*Ipomoea* spp.
Mullein, moth—*Verbascum blattaria* L.*
Muskmelon (cantaloup)—*Cucumis melo* L.
Mustard:
  Black—*Brassica nigra* (L.) Koch*
  India—*Brassica juncea* (L.) Czern.
Myosotis—*Myosotis* spp.
Nasturtium—*Tropaeolum majus* L.
Needlegrass—*Stipa viridula* Trin.*
Nemesia—*Nemesia* spp.
Nicotiana—*Nicotiana* spp.
Nigella—*Nigella damascena* L.
Nutgrass, nutsedge—*Cyperus rotundus* L.*
Oatgrass, tall—*Arrhenatherum elatius* (L.) Beauv. ex J. Presl and K. Presl
Oats—*Avena sativa* L.
Okra—*Hibiscus esculentus* L.
Onion—*Allium cepa* L.
Onion, Welch—*Allium fistulosum* L.
Orchardgrass—*Dactylis glomerata* L.
Pakchoi—*Brassica chinensis* L.
Panicgrass—*Panicum* spp.*
Panicgrass, blue—*Panicum antidotale* Retz
Pansy—*Viola tricolor* L.
Parsley—*Petroselinum crispum* (Mill.) Nym. ex A. W. Hill
Parsnip—*Pastinaca sativa* L.
Partridgepea—*Cassia fasciculata* Michx.*
Paspalum—*Paspalum* spp.*
Path rush—*Juncus bufonius* L.*
Pea:
  Austrian winter or field—*Pisum sativum* var. *arvense* (L.) Poir.
  Garden—*Pisum sativum* L.
  Rose or crown—*Pisum umbellatum* (*P. sativum* var. *umbellatum* Ser.)
Peanut, Florida runner—*Arachis hypogaea* L.
Pechay (mustard)—*Brassica juncea* (L.) Czern.
Pennisetum—*Pennisetum* spp.*
Pennycress, fanweed—*Thlaspi* spp.*
Penstemon—*Penstemon* spp.
Pepper—*Capsicum annuum* L.
Pepper, red—*Capsicum frutescens* L.
Pe-tsai (Chinese cabbage)—*Brassica pekinensis* (Lour.) Rupr.
Petunia—*Petunia hybrida* Vilm.
Phacelia—*Phacelia* spp.

Phlox—*Phlox drummondii* Hook. and *Phlox* spp.
Physalis—*Physalis* spp.
Pimpernel—*Anagallis* spp.*
Pine—*Pinus* spp.
Pinks, China—*Dianthus* spp.
Plantain—*Plantago* spp.*
Poppy:
  California—*Eschscholtzia californica* Cham.
  Corn, shirley—*Papaver rhoeas* L.
  Iceland—*Papaver nudicaule* L.
  Mexican tulip—*Hunnemannia fumariifolia* Sweet
  Oriental—*Papaver orientale* L.
Portulaca—*Portulaca grandiflora* Hook.
Potato—*Solanum tuberosum* L.
Primrose—*Primula* spp.
Primrose, Chinese—*Primula sinensis* Lindl.*
Pumpkin—*Cucurbita pepo* L. and *Cucurbita* spp.
Pyrethrum—*Chrysanthemum* spp.
Radish—*Raphanus sativus* L.
Radish, Japanese—*Raphanus sativus* L.
Rape, annual and winter—*Brassica napus* L.
Redtop—*Agrostis gigantea* Roth
Rescuegrass—*Bromus unioloides* Kunth
Rhodesgrass—*Chloris gayana* Kunth
Rhubarb—*Rheum rhaponticum* L.
Rice—*Oryza sativa* L.
Ricegrass, Indian—*Oryzopsis hymenoides* (Roem. and Schult.) Ricker
Rose campion—*Lychnis* spp.
Rosemary—*Rosmarinus officinalis* L.
Roughpea—*Lathyrus hirsutus* L.
Rush—*Juncus* spp.*
Rutabaga—*Brassica napus* var. *napobrassica* (L.) Reichb.
Rye—*Secale cereale* L.
Ryegrass:
  Annual—*Lolium multiflorum* Lam.
  Italian—*Lolium multiflorum* Lam.
  Perennial—*Lolium perenne* L.
Safflower—*Carthamus tinctorius* L.
Sage:
  Mealycup—*Salvia farinacea* Benth.
  Sage—*Salvia officinalis* L.
  Scarlet—*Salvia splendens* Sello ex Nees
Sainfoin—*Onobrychis viciifolia* Scop.
St.-Johnswort—*Hypericum humifusum* L.*

Salpiglossis—*Salpiglossis sinuata* Ruiz and Pav.
Salsify—*Tragopogon porrifolius* L.
Saponaria—*Saponaria ocymoides* L. and *S. vaccaria* L.
Savory—*Satureja hortensis* L.
Scabiosa—*Scabiosa atropurpurea* L. and *S. caucasica* Bieb.
Scotch-broom—*Cytisus candicans* Lam. (*C. monspessulanus*) and *C. scoparius* (L.) Link*
Senna—*Cassia bicapsularis* L. and *C. ultijuga* Rich.*
Sensitive plant—*Mimosa* spp.
Sesame—*Sesamum indicum* L.
Sesbania—*Sesbania* spp.
Silktree, mimosa—*Albizia julibrissin* Durazz.*
Sitao—*Vigna unguiculata* subsp. *sesquipedalis* (L.) Verdc.
Smartweed—*Polygonum* spp.*
Smartweed, water—*Polygonum hydropiper* L.*
Smilo—*Oryzopsis miliacea* (L.) Aschers. and Schweinf.
Snapdragon—*Antirrhinum* spp.
Sneezeweed, helenium—*Helenium* spp.
Snow-on-the-mountain—*Euphorbia marginata* Pursh
Solanum—*Solanum* spp.
Sorghum, grain and sweet—*Sorghum bicolor* (L.) Moench
Soybean—*Glycine max* (L.) Merr.
Spinach:
    Common—*Spinacia oleracea* L.
    New Zealand—*Tetragonia tetragonioides* (Pall.) Ktze.
Squash:
    Butternut—*Cucurbita pepo* L.
    Winter—*Cucurbita moschata* Duchesne ex Poir. and *C. maxima* Duchesne
Statice—*Statice* spp.
Stocks—*Matthiola* spp.
Strawberry—*Fragaria* × *ananassa* Duchesne
Strawflower—*Helichrysum monstrosum* Hort.
Sudangrass—*Sorghum sudanense* (Piper) Stapf
Sunflower—*Helianthus annuus* L.
Swede—*Brassica napus* var. *napobrassica* (L.) Reichb.
Sweetclover:
    White—*Melilotus alba* Desr.
    Yellow—*Melilotus officinalis* (L.) Pall.
Sweetpea:
    Annual—*Lathyrus odoratus* L.
    Perennial—*Lathyrus latifolius* L.
Sweet-william—*Dianthus barbatus* L.
Switchgrass—*Panicum virgatum* L.

Thyme—*Thymus vulgaris* L.
Timothy—*Phleum pratense* L.
Tobacco—*Nicotiana tabacum* L.
Tomato—*Lycopersicon esculentum* Mill.
Trefoil:
  Big—*Lotus uliginosus* Schkuhr
  Birdsfoot—*Lotus corniculatus* L.
Turnip—*Brassica rapa* L.
Vaseygrass—*Paspalum urvillei* Steud.
Velvetbean—*Mucuna deeringiana* (Bort) Merr.
Verbena—*Verbena* spp.
Vetch:
  Common—*Vicia sativa* L.
  Hairy—*Vicia villosa* Roth
  Hungarian—*Vicia pannonica* Crantz
  Monantha—*Vicia articulata* Hornem.
  Narrowleaf—*Vicia angustifolia* (L.) Reich.
  Purple—*Vicia benghalensis* L.
  Woollypod—*Vicia dasycarpa* Ten.
Vinca, periwinkle—*Vinca rosea* L. and *V. minor* L.
Viola—*Viola cornuta* L.
Wallflower—*Cheiranthus allioni* Hort. and *C. cheiri* L.
Walnut—*Juglans australis* Griseb.
Watermelon—*Citrullus lanatus* (Thunb.) Matsum. and Nakai
Wheat:
  Common—*Triticum aestivum* L.
  Durum—*Triticum durum* Desf.
  Hard red spring—*Triticum aestivum* L.
  Hard red winter—*Triticum aestivum* L.
  Poulard—*Triticum turgidum* L.*
  Red Winter Speltz—*Triticum spelta* L.
  Soft—*Triticum* spp.
  Soft red winter—*Triticum aestivum* L.
  Spring—*Triticum aestivum* L.
  White—*Triticum* spp.
Wheatgrass:
  Fairway crested—*Agropyron cristatum* (L.) Gaertn.
  Intermediate—*Agropyron intermedium* (Host) Beauv.
  Pubescent—*Agropyron trichophorum* (Link) Richt.

Wheatgrass—Continued

Slender—*Agropyron trachycaulum* (Link) Malte ex H. F. Lewis
Standard crested—*Agropyron desertorum* (Fisch. ex Link) Schult.
Tall—*Agropyron elongatum* (Host) Beauv.
Western—*Agropyron smithii* Rydb.

Wild rye:

Blue—*Elymus glaucus* Buckl.*
Canada—*Elymus canadensis* L.
Russian—*Elymus junceus* Fisch.

Willow, weeping—*Salix babylonica* L.
Zinnia—*Zinnia* spp.
Zoysia (see Japanese lawngrass and manilagrass)

Bromus polyanthus Scribn.*
Dioclea pauciflora Rusby*
Hovea linearis (J. E. Smith) R. Br.*
Kennedya apetala Loddiger*
Medicago orbicularis (L.) Bartal.*
Melilotus gracilis DC.*

Insects

Beetle:

Flat grain—*Cryptolestes pusillus* (Schönherr)
Khapra—*Trogoderma granarium* Everts
Sawtoothed grain—*Oryzaephilus surinamensis* (L.)

Borer, lesser grain—*Rhyzopertha dominica* (F.)
Cadelle—*Tenebroides mauritanicus* (L.)
Chalcid—*Bruchophagus* spp.
Grain moth, Angoumois—*Sitotroga cerealella* (Oliver)
Weevil:

Granary—*Sitophilus granarius* (L.)
Rice—*Sitophilus oryzae* (L.)

Animals

Lemming, collard—*Dicrostonyx* spp.
Mouse—*Muridae* spp.
Rat—*Rattus* spp.
Squirrel—*Sciurus* spp.

CONVERSION TABLES FOR TEMPERATURES AND MEASURES

Temperature Conversion[12]

The numbers in the center of each column refer to the temperature in centigrade or Fahrenheit. They can be converted to either scale. If converting Fahrenheit to centigrade, the equivalent temperature is on the left in each column. If converting centigrade to Fahrenheit, the equivalent temperature is on the right.

Degrees C	Degrees	Degrees F	Degrees C	Degrees	Degrees F	Degrees C	Degrees	Degrees F
-40.0	**-40**	-40.0	-24.4	**-12**	+10.4	-8.9	**+16**	+60.8
			-23.9	**-11**	+12.2	-8.3	**+17**	+62.6
-39.4	**-39**	-38.2	-23.3	**-10**	+14.0	-7.8	**+18**	+64.4
-38.9	**-38**	-36.4				-7.2	**+19**	+66.2
-38.3	**-37**	-34.6	-22.8	**-9**	+15.8	-6.7	**+20**	+68.0
-37.8	**-36**	-32.8	-22.2	**-8**	+17.6			
-37.2	**-35**	-31.0	-21.7	**-7**	+19.4	-6.1	**+21**	+69.8
			-21.1	**-6**	+21.2	-5.5	**+22**	+71.6
-36.7	**-34**	-29.2	-20.6	**-5**	+23.0	-5.0	**+23**	+73.4
-36.1	**-33**	-27.4				-4.4	**+24**	+75.2
-35.6	**-32**	-25.6	-20.0	**-4**	+24.8	-3.9	**+25**	+77.0
-35.0	**-31**	-23.8	-19.4	**-3**	+26.6			
-34.4	**-30**	-22.0	-18.9	**-2**	+28.4	-3.3	**+26**	+78.8
			-18.3	**-1**	+30.2	-2.8	**+27**	+80.6
-33.9	**-29**	-20.2	-17.8	**0**	+32.0	-2.2	**+28**	+82.4
-33.3	**-28**	-18.4				-1.7	**+29**	+84.2
-32.8	**-27**	-16.6	-17.2	**+1**	+33.8	-1.1	**+30**	+86.0
-32.2	**-26**	-14.8	-16.7	**+2**	+35.6			
-31.7	**-25**	-13.0	-16.1	**+3**	+37.4	-.6	**+31**	+87.8
			-15.6	**+4**	+39.2	.0	**+32**	+89.6
-31.1	**-24**	-11.2	-15.0	**+5**	+41.0	+.6	**+33**	+91.4
-30.6	**-23**	-9.4				+1.1	**+34**	+93.2
-30.0	**-22**	-7.6	-14.4	**+6**	+42.8	+1.7	**+35**	+95.0
-29.4	**-21**	-5.8	-13.9	**+7**	+44.6			
-28.9	**-20**	-4.0	-13.3	**+8**	+46.4	+2.2	**+36**	+96.8
			-12.8	**+9**	+48.2	+2.8	**+37**	+98.6
-28.3	**-19**	-2.2	-12.2	**+10**	+50.0	+3.3	**+38**	+100.4
-27.8	**-18**	-0.4				+3.9	**+39**	+102.2
-27.2	**-17**	+1.4	-11.7	**+11**	+51.8	+4.4	**+40**	+104.0
-26.7	**-16**	+3.2	-11.1	**+12**	+53.6			
-26.1	**-15**	+5.0	-10.6	**+13**	+55.4	+5.0	**+41**	+105.8
			-10.0	**+14**	+57.2	+5.5	**+42**	+107.6
-25.6	**-14**	+6.8	-9.4	**+15**	+59.0	+6.1	**+43**	+109.4
-25.0	**-13**	+8.6				+6.7	**+44**	+111.2

[12] Tabular data courtesy of Weksler Thermometer Corp., Freeport, Long Island, N.Y. Absolute zero is -273.16° C or -459.69° F.

Degrees C	Degrees	Degrees F
+7.2	**+45**	+113.0
+7.8	**+46**	+114.8
+8.3	**+47**	+116.6
+8.9	**+48**	+118.4
+9.4	**+49**	+120.2
+10.0	**+50**	+122.0
+10.6	**+51**	+123.8
+11.1	**+52**	+125.6
+11.7	**+53**	+127.4
+12.2	**+54**	+129.2
+12.8	**+55**	+131.0
+13.3	**+56**	+132.8
+13.9	**+57**	+134.6
+14.4	**+58**	+136.4
+15.0	**+59**	+138.2
+15.6	**+60**	+140.0
+16.1	**+61**	+141.8
+16.7	**+62**	+143.6
+17.2	**+63**	+145.4
+17.8	**+64**	+147.2
+18.3	**+65**	+149.0
+18.9	**+66**	+150.8
+19.4	**+67**	+152.6
+20.0	**+68**	+154.4
+20.6	**+69**	+156.2
+21.1	**+70**	+158.0
+21.7	**+71**	+159.8
+22.2	**+72**	+161.6
+22.8	**+73**	+163.4
+23.3	**+74**	+165.2
+23.9	**+75**	+167.0
+24.4	**+76**	+168.8
+25.0	**+77**	+170.6
+25.6	**+78**	+172.4
+26.1	**+79**	+174.2
+26.7	**+80**	+176.0
+27.2	**+81**	+177.8
+27.8	**+82**	+179.6
+28.3	**+83**	+181.4
+28.9	**+84**	+183.2
+29.4	**+85**	+185.0

Degrees C	Degrees	Degrees F
+30.0	**+86**	+186.8
+30.6	**+87**	+188.6
+31.1	**+88**	+190.4
+31.7	**+89**	+192.2
+32.2	**+90**	+194.0
+32.8	**+91**	+195.8
+33.3	**+92**	+197.6
+33.9	**+93**	+199.4
+34.4	**+94**	+201.2
+35.0	**+95**	+203.0
+35.6	**+96**	+204.8
+36.1	**+97**	+206.6
+36.7	**+98**	+208.4
+37.2	**+99**	+210.2
+37.8	**+100**	+212.0
+38.3	**+101**	+213.8
+38.9	**+102**	+215.6
+39.4	**+103**	+217.4
+40.0	**+104**	+219.2
+40.6	**+105**	+221.0
+41.1	**+106**	+222.8
+41.7	**+107**	+224.6
+42.2	**+108**	+226.4
+42.8	**+109**	+228.2
+43.3	**+110**	+230.0
+43.9	**+111**	+231.8
+44.4	**+112**	+233.6
+45.0	**+113**	+235.4
+45.6	**+114**	+237.2
+46.1	**+115**	+239.0
+46.7	**+116**	+240.8
+47.2	**+117**	+242.6
+47.8	**+118**	+244.4
+48.3	**+119**	+246.2
+48.9	**+120**	+248.0
+49.4	**+121**	+249.8
+50.0	**+122**	+251.6
+50.6	**+123**	+253.4
+51.1	**+124**	+255.2
+51.7	**+125**	+257.0
+52.2	**+126**	+258.8
+52.8	**+127**	+260.6

Degrees C	Degrees	Degrees F
+53.3	**+128**	+262.4
+53.9	**+129**	+264.2
+54.4	**+130**	+266.0
+55.0	**+131**	+267.8
+55.6	**+132**	+269.6
+56.1	**+133**	+271.4
+56.7	**+134**	+273.2
+57.2	**+135**	+275.0
+57.8	**+136**	+276.8
+58.3	**+137**	+278.6
+58.9	**+138**	+280.4
+59.4	**+139**	+282.2
+60.0	**+140**	+284.0
+60.6	**+141**	+285.8
+61.1	**+142**	+287.6
+61.7	**+143**	+289.4
+62.2	**+144**	+291.2
+62.8	**+145**	+293.0
+63.3	**+146**	+294.8
+63.9	**+147**	+296.6
+64.4	**+148**	+298.4
+65.0	**+149**	+300.2
+65.6	**+150**	+302.0
+66.1	**+151**	+303.8
+66.7	**+152**	+305.6
+67.2	**+153**	+307.4
+67.8	**+154**	+309.2
+68.3	**+155**	+311.0
+68.9	**+156**	+312.8
+69.4	**+157**	+314.6
+70.0	**+158**	+316.4
+70.6	**+159**	+318.2
+71.1	**+160**	+320.0
+71.7	**+161**	+321.8
+72.2	**+162**	+323.6
+72.8	**+163**	+325.4
+73.3	**+164**	+327.2
+73.9	**+165**	+329.0
+74.4	**+166**	+330.8
+75.0	**+167**	+332.6
+75.6	**+168**	+334.4

Degrees C	Degrees	Degrees F	Degrees C	Degrees	Degrees F	Degrees C	Degrees	Degrees F
+76.1	**+169**	+336.2	+95.6	**+204**	+399.2	+115.0	**+239**	+462.2
+76.7	**+170**	+338.0	+96.1	**+205**	+401.0	+115.6	**+240**	+464.0
+77.2	**+171**	+339.8	+96.7	**+206**	+402.8	+116.1	**+241**	+465.8
+77.8	**+172**	+341.6	+97.2	**+207**	+404.6	+116.7	**+242**	+467.6
+78.3	**+173**	+343.4	+97.8	**+208**	+406.4	+117.2	**+243**	+469.4
+78.9	**+174**	+345.2	+98.3	**+209**	+408.2	+117.8	**+244**	+471.2
+79.4	**+175**	+347.0	+98.9	**+210**	+410.0	+118.3	**+245**	+473.0
+80.0	**+176**	+348.8	+99.4	**+211**	+411.8	+118.9	**+246**	+474.8
+80.6	**+177**	+350.6	+100.0	**+212**	+413.6	+119.4	**+247**	+476.6
+81.1	**+178**	+352.4	+100.6	**+213**	+415.4	+120.0	**+248**	+478.4
+81.7	**+179**	+354.2	+101.1	**+214**	+417.2	+120.6	**+249**	+480.2
+82.2	**+180**	+356.0	+101.7	**+215**	+419.0	+121.1	**+250**	+482.0
+82.8	**+181**	+357.8	+102.2	**+216**	+420.8	+122.4	**+252**	+485.6
+83.3	**+182**	+359.6	+102.8	**+217**	+422.6	+123.3	**+254**	+489.2
+83.9	**+183**	+361.4	+103.3	**+218**	+424.4	+124.4	**+256**	+492.8
+84.4	**+184**	+363.2	+103.9	**+219**	+426.2	+125.5	**+258**	+496.4
+85.0	**+185**	+365.0	+104.4	**+220**	+428.0	+126.7	**+260**	+500.0
+85.6	**+186**	+366.8	+105.0	**+221**	+429.8	+127.8	**+262**	+503.6
+86.1	**+187**	+368.6	+105.6	**+222**	+431.6	+128.9	**+264**	+507.2
+86.7	**+188**	+370.4	+106.1	**+223**	+433.4	+130.0	**+266**	+510.8
+87.2	**+189**	+372.2	+106.7	**+224**	+435.2	+131.3	**+268**	+514.4
+87.8	**+190**	+374.0	+107.2	**+225**	+437.0	+132.2	**+270**	+518.0
+88.3	**+191**	+375.8	+107.8	**+226**	+438.8	+133.3	**+272**	+521.6
+88.9	**+192**	+377.6	+108.3	**+227**	+440.6	+134.4	**+274**	+525.2
+89.4	**+193**	+379.4	+108.9	**+228**	+442.4	+135.6	**+276**	+528.8
+90.0	**+194**	+381.2	+109.4	**+229**	+444.2	+136.7	**+278**	+532.4
+90.6	**+195**	+383.0	+110.0	**+230**	+446.0	+137.8	**+280**	+536.0
+91.1	**+196**	+384.8	+110.6	**+231**	+447.8	+138.9	**+282**	+539.6
+91.7	**+197**	+386.6	+111.1	**+232**	+449.6	+140.0	**+284**	+543.2
+92.2	**+198**	+388.4	+111.7	**+233**	+451.4	+141.1	**+286**	+546.8
+92.8	**+199**	+390.2	+112.2	**+234**	+453.2	+142.2	**+288**	+550.4
+93.3	**+200**	+392.0	+112.8	**+235**	+455.0	+143.3	**+290**	+554.0
+93.9	**+201**	+393.8	+113.3	**+236**	+456.8	+144.4	**+292**	+557.6
+94.4	**+202**	+395.6	+113.9	**+237**	+458.6	+145.6	**+294**	+561.2
+95.0	**+203**	+397.4	+114.4	**+238**	+460.4	+146.7	**+296**	+564.8
						+147.8	**+298**	+568.4

Measures of Weight

Avoirdupois to metric:

1 ounce (oz)	28.3 grams
1 pound (lb)	453.6 grams
1 hundredweight (cwt)	50.8 kilograms
1 short ton (2,000 lb)	907.0 kilograms
1 long ton (2,240 lb)	1,016.0 kilograms
1 metric ton (1 tonne)	1,000.0 kilograms

Metric to avoirdupois:

1 gram (gm)	0.3533 ounce
1 kilogram (kg)	2.205 pounds
1 metric ton (1,000 kg)	1.102 short tons

Measures of Length

British or American to metric:

1 inch (in)	2.54 centimeters
1 foot (ft)	30.48 centimeters
1 yard (yd)	.914 meter
1 mile (mi)	1.609 kilometers

Metric to British or American:

1 millimeter (mm)	0.039 inch
1 centimeter (cm)	.394 inch
1 decimeter (dm)	3.397 inches
1 meter (m)	1.094 yards
1 kilometer (km)	.621 mile

Measures of Area

British or American to metric:

1 square inch (in^2)	6.452 square centimeters
1 square foot (ft^2)	9.29 square decimeters
1 square yard (yd^2)	.836 square meter
1 acre	.405 hectare

Metric to British or American:

1 square millimeter (mm^2)	0.155 square inch
1 square meter (m^2)	1.196 square yards
1 hectare (ha)	2.471 acres
1 square kilometer (km^2)	.386 square mile

Measures of Volume

British or American to metric:

1 cubic inch (in^3) .... 16.387 cubic centimeters
1 cubic foot (ft^3) .... 28.317 cubic decimeters
1 cubic yard (yd^3) .... .765 cubic meter

Metric to British or American:

1 cubic centimeter (cm^3) .... 0.061 cubic inch
1 cubic decimeter (dm^3) .... .035 cubic foot
1 cubic meter (m^3) .... 1.308 cubic yards

Measures of Capacity

American and British to metric:

1 pint (pt) (American) .... 0.473 liter
1 quart (qt) (American) .... .946 liter
1 gallon (gal) (American) .... 3.785 liters
1 gallon (British) .... 4.546 liters

Metric to American and British:

1 liter (1) .... 1.057 quarts (American)
1 liter .... .880 quart (British)

LITERATURE CITED

ANONYMOUS.
1942a. SEED STORAGE PROBLEMS. Puerto Rico Agr. Expt. Sta. Ann. Rpt. 1941/42: 39.

———
1942b. DURATION OF VIABILITY OF SEEDS. Gard. Chron. 111: 234.

———
1959. SEEDS BAGGED TO KEEP. Agr. Res. 8: 10–11.

———
1968. 550-YEAR OLD SEED SPROUTS. Sci. News 94: 367.

ABDALLA, F. H., and ROBERTS, E. H.
1969a. THE EFFECTS OF TEMPERATURE AND MOISTURE ON THE INDUCTION OF GENETIC CHANGES IN SEEDS OF BARLEY, BROAD BEANS, AND PEAS DURING STORAGE. Ann. Bot. (N.S.) 33: 153–167.

——— and ROBERTS, E. H.
1969b. THE EFFECT OF SEED STORAGE CONDITIONS ON THE GROWTH AND YIELD OF BARLEY, BROAD BEANS, AND PEAS. Ann. Bot. (N.S.) 33: 169–184.

ABDUL-BAKI, A. A., and ANDERSON, J. D.
1972. PHYSIOLOGICAL AND BIOCHEMICAL DETERIORATION OF SEEDS. *In* Kozlowski, T. T., Seed Biology, v. 2, pp. 283–315, illus. New York.

AGENA, M. U.
1961. UNTERSUCHUNGEN UBER KALTEEINWIRKUNGEN AUF LAGERNDE GETRIEDEFRUCHTE MIT VERSCHIEDENEM WASSERGEHALT. 112 pp. Bonn.

ALTSCHUL, A. M., KARON, M. L., KAYME, L., and HALL, C. M.
1946. EFFECT OF INHIBITORS ON THE RESPIRATION AND STORAGE OF COTTONSEED. Plant Physiol. 21: 573–587.

American Society of Agricultural Engineers.
1972. ASAE DATA: ASAE D245.1. MOISTURE RELATIONSHIPS OF GRAIN. Agr. Engin. Ybk. 1971, pp. 364–366.

Anderson, J. A., and Alcock, A. W., eds.
1954. STORAGE OF CEREAL GRAINS AND THEIR PRODUCTS. 515 pp., illus. St. Paul, Minn.

Arndt, C. H.
1946. THE INTERNAL INFECTION OF COTTON SEED AND THE LOSS OF VIABILITY IN STORAGE. Phytopathology 36: 30–37.

Arnold, J. R., and Libby, W. F.
1951. RADIOCARBON DATES. Science 113: 111–120.

Associated Seed Growers, Inc. (Asgrow Seed Co.).
1949. A STUDY OF MECHANICAL INJURY TO SEED BEANS. Asgrow Monog. 1, 46 pp.

———
1954. THE PRESERVATION OF VIABILITY AND VIGOR IN VEGETABLE SEED. Asgrow Monog. 2, 32 pp.

Association of Official Seed Analysts.
1954. RULES FOR TESTING SEEDS. Assoc. Off. Seed Anal. Proc. 44: 31–78.

———
1970. RULES FOR TESTING SEEDS. Assoc. Off. Seed Anal. Proc. 60: 1–116.

Atkin, J. D.
1958. RELATIVE SUSCEPTIBILITY OF SNAP BEAN VARIETIES TO MECHANICAL INJURY OF SEED. Amer. Soc. Hort. Sci. Proc. 72: 370–373.

Atkins, W. R. G.
1909. THE ABSORPTION OF WATER BY SEEDS. Roy. Dublin Soc. Sci. Proc. 12: 35–46.

Aufhammer, G., and Simon, U.
1957. DIE SAMEN LANDWIRTSHAFTLICHER KULTURPFLANZEN IM GRUNDSTEIN DES CHAMALIGEN NURNBERGER STADTTHEATERS UND IHRE KEIMFAHIGKEIT. Ztschr. f. Acker- u. Pflanzenbau. 103: 454–472.

Austin, R. B.
1972. EFFECTS OF ENVIRONMENT BEFORE HARVESTING ON VIABILITY. *In* Roberts, E. H., Viability of Seeds, pp. 114–149, illus. London.

Avery, A. G., and Blakeslee, A. F.
1936. VISIBLE MUTATIONS FROM AGED SEEDS. Amer. Nat. 70: 36–37.

Babbitt, J. D.
1949. OBSERVATIONS ON THE ADSORPTION OF WATER VAPOR BY WHEAT. Canad. Jour. Res., Sect. F, Technol. 27: 55–72.

Bacchi, O.
1955. SECA DA SEMENTE DE CAFE AO SOL. Bragantia 14: 225–236.

Bailey, C. H.
1940. RESPIRATION OF CEREAL GRAINS AND FLAXSEED. Plant Physiol. 15: 257–274.

Bainer, R., and Borthwick, H. A.
1934. THRESHER AND OTHER MECHANICAL INJURY TO SEED BEANS OF THE LIMA TYPE. Calif. Bul. 580, 30 pp.

Baker, D., Neustadt, M. H., and Zeleny, L.
1957. APPLICATION OF THE FAT ACIDITY TEST AS AN INDEX OF GRAIN DETERIORATION. Cereal Chem. 34: 226–233.

Baker, K. F.
1972. SEED PATHOLOGY. *In* Kozlowski, T. T., Seed Biology, v. 2, pp. 317–416, illus. New York.

Barre, H. J.

1954. COUNTRY STORAGE OF GRAIN. *In* Anderson, J. A., and Alcock, A. W., eds., Storage of Cereal Grains and Their Products, pp. 308–357. St. Paul, Minn.

——— 1963. IMPORTANT EQUIPMENT FOR DRYING SEEDS IN NORTH AMERICA. Internatl. Seed Testing Assoc. Proc. 28: 815–826.

Barrow, E. H. R.

1915. THE HEATING OF COTTONSEED—ITS CAUSES AND PREVENTION. Indus. Engin. Chem. 7: 709–712.

Bartholomew, D. P., and Loomis, W. E.

1967. CARBON DIOXIDE PRODUCTION BY DRY GRAIN IN ZEA MAYS. Plant Physiol. 42: 120–124.

Barton, L. V.

1932. EFFECT OF STORAGE ON THE VITALITY OF DELPHINIUM SEEDS. Boyce Thompson Inst. Contrib. 4: 141–153.

——— 1939. STORAGE OF SOME FLOWER SEEDS. Boyce Thompson Inst. Contrib. 10: 399–427.

——— 1941. RELATION OF CERTAIN AIR TEMPERATURES AND HUMIDITIES TO VIABILITY OF SEEDS. Boyce Thompson Inst. Contrib. 12: 85–102.

——— 1947. EFFECT OF DIFFERENT STORAGE CONDITIONS ON THE GERMINATION OF SEEDS OF CINCHONA LEDGERIANA MOENS. Boyce Thompson Inst. Contrib. 15: 1–10.

——— 1949. SEED PACKETS AND ONION SEED VIABILITY. Boyce Thompson Inst. Contrib. 15: 341–352.

——— 1953. SEED STORAGE AND VIABILITY. Boyce Thompson Inst. Contrib. 17: 87–103.

——— 1954. EFFECT OF SUBFREEZING TEMPERATURES ON VIABILITY OF CONIFER SEEDS IN STORAGE. Boyce Thompson Inst. Contrib. 18: 21–24.

——— 1960a. STORAGE OF SEEDS OF LOBELIA CARDINALIS L. Boyce Thompson Inst. Contrib. 20: 395–401.

——— 1960b. LIFE SPAN OF FROST-DAMAGED CORN SEEDS. Boyce Thompson Inst. Contrib. 20: 403–408.

——— 1961. SEED PRESERVATION AND LONGEVITY. 216 pp., illus. London and New York.

——— 1966a. THE EFFECT OF STORAGE CONDITIONS ON THE VIABILITY OF BEAN SEEDS. Boyce Thompson Inst. Contrib. 23: 281–284.

——— 1966b. VIABILITY OF PYRETHRUM SEEDS. Boyce Thompson Inst. Contrib. 23: 267–268.

——— and Garman, H. R.

1946. EFFECT OF AGE AND STORAGE CONDITION OF SEEDS ON THE YIELDS OF CERTAIN PLANTS. Boyce Thompson Inst. Contrib. 14: 243–255.

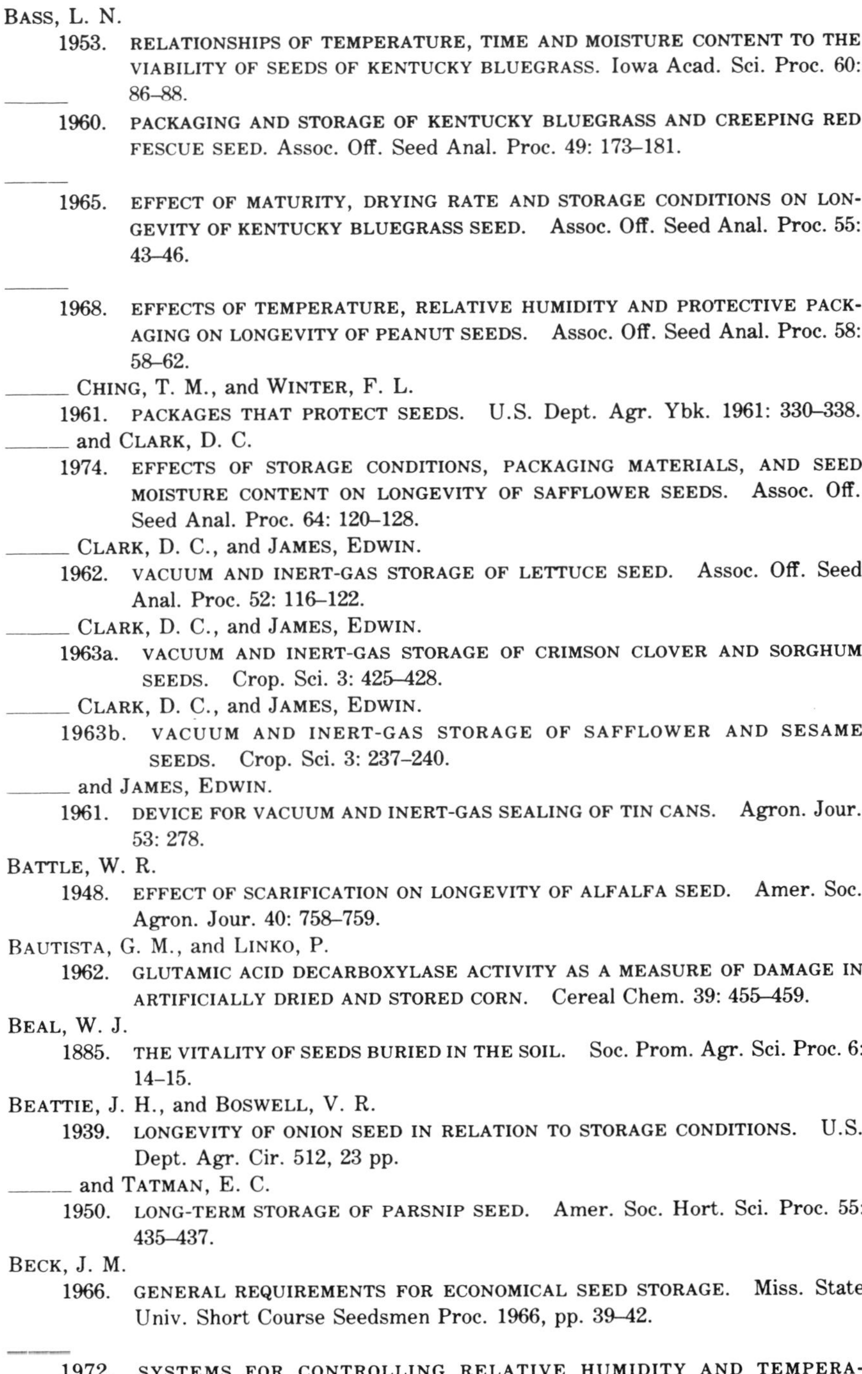

BASS, L. N.
1953. RELATIONSHIPS OF TEMPERATURE, TIME AND MOISTURE CONTENT TO THE VIABILITY OF SEEDS OF KENTUCKY BLUEGRASS. Iowa Acad. Sci. Proc. 60: 86–88.

______ 1960. PACKAGING AND STORAGE OF KENTUCKY BLUEGRASS AND CREEPING RED FESCUE SEED. Assoc. Off. Seed Anal. Proc. 49: 173–181.

______ 1965. EFFECT OF MATURITY, DRYING RATE AND STORAGE CONDITIONS ON LONGEVITY OF KENTUCKY BLUEGRASS SEED. Assoc. Off. Seed Anal. Proc. 55: 43–46.

______ 1968. EFFECTS OF TEMPERATURE, RELATIVE HUMIDITY AND PROTECTIVE PACKAGING ON LONGEVITY OF PEANUT SEEDS. Assoc. Off. Seed Anal. Proc. 58: 58–62.

______ CHING, T. M., and WINTER, F. L.
1961. PACKAGES THAT PROTECT SEEDS. U.S. Dept. Agr. Ybk. 1961: 330–338.

______ and CLARK, D. C.
1974. EFFECTS OF STORAGE CONDITIONS, PACKAGING MATERIALS, AND SEED MOISTURE CONTENT ON LONGEVITY OF SAFFLOWER SEEDS. Assoc. Off. Seed Anal. Proc. 64: 120–128.

______ CLARK, D. C., and JAMES, EDWIN.
1962. VACUUM AND INERT-GAS STORAGE OF LETTUCE SEED. Assoc. Off. Seed Anal. Proc. 52: 116–122.

______ CLARK, D. C., and JAMES, EDWIN.
1963a. VACUUM AND INERT-GAS STORAGE OF CRIMSON CLOVER AND SORGHUM SEEDS. Crop. Sci. 3: 425–428.

______ CLARK, D. C., and JAMES, EDWIN.
1963b. VACUUM AND INERT-GAS STORAGE OF SAFFLOWER AND SESAME SEEDS. Crop. Sci. 3: 237–240.

______ and JAMES, EDWIN.
1961. DEVICE FOR VACUUM AND INERT-GAS SEALING OF TIN CANS. Agron. Jour. 53: 278.

BATTLE, W. R.
1948. EFFECT OF SCARIFICATION ON LONGEVITY OF ALFALFA SEED. Amer. Soc. Agron. Jour. 40: 758–759.

BAUTISTA, G. M., and LINKO, P.
1962. GLUTAMIC ACID DECARBOXYLASE ACTIVITY AS A MEASURE OF DAMAGE IN ARTIFICIALLY DRIED AND STORED CORN. Cereal Chem. 39: 455–459.

BEAL, W. J.
1885. THE VITALITY OF SEEDS BURIED IN THE SOIL. Soc. Prom. Agr. Sci. Proc. 6: 14–15.

BEATTIE, J. H., and BOSWELL, V. R.
1939. LONGEVITY OF ONION SEED IN RELATION TO STORAGE CONDITIONS. U.S. Dept. Agr. Cir. 512, 23 pp.

______ and TATMAN, E. C.
1950. LONG-TERM STORAGE OF PARSNIP SEED. Amer. Soc. Hort. Sci. Proc. 55: 435–437.

BECK, J. M.
1966. GENERAL REQUIREMENTS FOR ECONOMICAL SEED STORAGE. Miss. State Univ. Short Course Seedsmen Proc. 1966, pp. 39–42.

______ 1972. SYSTEMS FOR CONTROLLING RELATIVE HUMIDITY AND TEMPERATURE. Miss. State Univ. Short Course Seedsmen Proc. 15: 77–85.

BECQUEREL, PAUL.

1907. RECHERCHES SUR LA VIE TATENTE DES GRAINES. Ann. des Sci. Nat., Bot. 5/6: 193–311.

——— 1934. LA LONGEVITE DES GRAINES MACROBIOTIQUES. [Paris] Acad. des. Sci. Compt. Rend. 199: 1662–1664.

——— 1953. LA SUSPENSION DE LA VIE AUX CONFINS DU ZERO ABSOLUET SES CONSEQUENCES. Extrait des Actes due Congrès de Luxembourg. 72d Sess. Assoc. Franç. pour l'Avanc. des Sci., pp. 487–491.

BLACK, J. N.

1959. SEED SIZE IN HERBAGE LEGUMES. Herbage Abs. 29: 235–241.

BLACKSTONE, J. H., WARD, H. S., JR., BUTT, J. L., and others.

1954. FACTORS AFFECTING GERMINATION OF RUNNER PEANUTS. Ala. Agr. Expt. Sta. Bul. 289, 31 pp.

BLAKESLEE, A. F.

1954. THE AGING OF SEEDS AND MUTATION RATES. N.Y. Acad. Sci. Ann. 57: 488–490.

——— AVERY, A. G., BERGNER, A. D., and SATINA, S.

1942. DATURA STUDIES. Carnegie Inst. Wash. Ybk. 41 (1941–42): 176–180.

BODGER SEEDS LTD.

1932. SOME INTERESTING STATISTICS ON THE LONGEVITY OF FLOWER SEEDS. Seed World 31 (11): 17.

BONVICINI, M.

1951. SEMENTE DI GRANO SENZA LESIONI. Commun. Cong. Assoc. Franç. Prog. Sci., p. i.

BORTHWICK, H. A.

1932. THRESHER INJURY IN BABY LIMA BEANS. Jour. Agr. Res. 44: 503–510.

BOSWELL, V. R., TOOLE, E. H., TOOLE, V. K., and FISHER, D. F.

1940. A STUDY OF RAPID DETERIORATION OF VEGETABLE SEEDS AND METHODS FOR ITS PREVENTION. U.S. Dept. Agr. Tech. Bul. 708, 40 pp.

BRANDENBURG, N. R., SIMONS, J. W., and SMITH, L. L.

1961. WHY AND HOW SEEDS ARE DRIED. U.S. Dept. Agr. Ybk. 1961, pp. 295–306.

BREESE, M. H.

1955. HYSTERESIS IN THE HYGROSCOPIC EQUILIBRIA OF ROUGH RICE AT 25° C. Cereal Chem. 32: 481–487.

BRETT, C. C.

1952a. FACTORS AFFECTING THE VIABILITY OF GRASS AND LEGUME SEED IN STORAGE AND DURING SHIPMENT. Internatl. Grassland Conf. 6: 878–884.

——— 1952b. THE INFLUENCE OF STORAGE CONDITIONS UPON THE LONGEVITY OF SEEDS, WITH SPECIAL REFERENCE TO ROOT AND VEGETABLE CROPS. 13th Internatl. Hort. Cong. Rpt., pp. 1016–1018.

BREWER, H. E., and BUTT, J. L.

1950. HYGROSCOPIC EQUILIBRIUM AND VIABILITY OF NATURALLY AND ARTIFICIALLY DRIED BLUE LUPINE SEEDS. Plant Physiol. 25: 245–268.

BRISON, F. R.

1941. INFLUENCE OF STORAGE CONDITIONS UPON THE GERMINATION OF ONION SEED. Tex. Acad. Sci. Proc. and Trans. 25: 69–70.

——— 1942. THE INFLUENCE OF STORAGE CONDITIONS UPON THE GERMINATION OF ONION SEED. Amer. Soc. Hort. Sci. Proc. 40: 501–503.

Brooke, C. L.
1935. MUMMY WHEAT WON'T GROW. Northwest. Miller 181: 456.

Brown, C. W., and Myers, A.
1960. LONG TERM CEREAL STORAGE. Agr. Gaz. N.S. Wales 71: 348–355.

Brown, E.
1939. PRESERVING THE VIABILITY OF BERMUDA ONION SEED. Science 89: 292–293.

_____ Stanton, T. R., Wiebe, G. A., and Martin, J. H.
1948. DORMANCY AND THE EFFECT OF STORAGE ON OATS, BARLEY, AND SORGHUM. U.S. Dept. Agr. Tech. Bul. 953, 30 pp.

Brown, E. B.
1922. EFFECTS OF MUTILATING THE SEED ON THE GROWTH AND PRODUCTIVENESS OF CORN. U.S. Dept. Agr. Bul. 1011, 14 pp.

Brown, H. T., and Escombe, F.
1897–98. NOTE ON THE INFLUENCE OF VERY LOW TEMPERATURES ON THE GERMINATIVE POWER OF SEEDS. Roy. Soc. London, Proc., Sect. B, 62: 160–165.

Bunch, H. D.
1958. MORE RESEARCH, EDUCATION, "PREACHING" NEEDED TO CUT DOWN ON HIGH SEED MOISTURE. Seedsmen's Digest 9 (5): 13.

Bunting, R. H.
1930. PROCEEDINGS OF THE ASSOCIATION OF ECONOMIC BIOLOGISTS. II. MYCOLOGICAL ASPECTS. Ann. Appl. Biol. 17: 402–407.

Burgess, J. L.
1938. REPORT ON PROJECT TO DETERMINE THE PERCENTAGE AND DURATION OF VIABILITY OF DIFFERENT VARIETIES OF SOYBEANS GROWN IN NORTH CAROLINA. Assoc. Off. Seed Anal. Proc. 23: 69.

Burlison, W. L., Van Doren, C. A., and Hackleman, J. C.
1940. ELEVEN YEARS OF SOYBEAN INVESTIGATIONS. VARIETIES, SEEDING, STORAGE. Ill. Agr. Expt. Sta. Bul. 462, pp. 121–167.

Burns, R. E.
1957. EFFECT OF AGE OF SEED ON SURVIVAL UNDER ADVERSE STORAGE CONDITIONS. (Abstract) Assoc. South. Agr. Workers Proc. 54: 226.

Caldwell, W. P.
1962. SAFER SEED STORAGE FOR SHIPMENT ABROAD. Crops and Soils 61–62: 17.

_____ and Bunch, H. D.
1961. EFFECTS OF SEED MOISTURE AND PACKAGING CONTAINER UPON VIABILITY AND VIGOR OF WHEAT, CABBAGE AND SOYBEAN SEEDS DURING TRANSOCEANIC SHIPMENT AND SUBSEQUENT STORAGE. Agron. Abs. 1961: 72.

California Department of Agriculture.
1973. CALIFORNIA SEED LAW AND REGULATIONS. 46 pp. Sacramento, Calif.

Carlson, G. E., and Atkins, R. E.
1960. EFFECT OF FREEZING TEMPERATURES ON SEED VIABILITY AND SEEDLING VIGOR OF GRAIN SORGHUM. Agron. Jour. 52: 329–333.

Chaney, R. W.
1951. HOW OLD ARE THE MANCHURIAN LOTUS SEEDS? Gard. Jour. N.Y. Bot. Gard. 1: 137–139.

Ching, T. M.
1959. KEEPING FIELD SEED QUALITY IN STORAGE AND TRANSIT. West. Feed and Seed 14: 21–22.

_____
1961. RESPIRATION OF FORAGE SEED IN HERMETICALLY SEALED CANS. Agron. Jour. 53: 6–8.

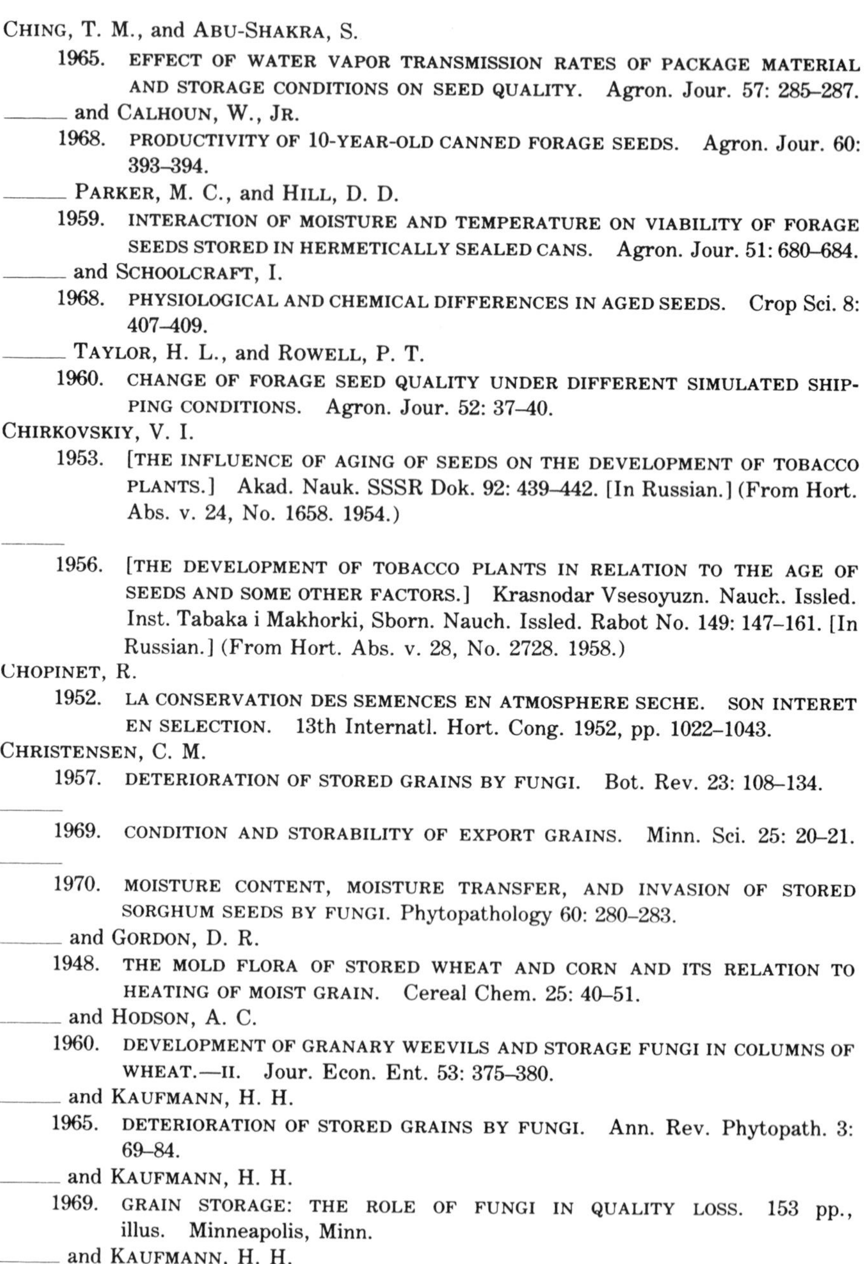

Ching, T. M., and Abu-Shakra, S.
1965. EFFECT OF WATER VAPOR TRANSMISSION RATES OF PACKAGE MATERIAL AND STORAGE CONDITIONS ON SEED QUALITY. Agron. Jour. 57: 285–287.

——— and Calhoun, W., Jr.
1968. PRODUCTIVITY OF 10-YEAR-OLD CANNED FORAGE SEEDS. Agron. Jour. 60: 393–394.

——— Parker, M. C., and Hill, D. D.
1959. INTERACTION OF MOISTURE AND TEMPERATURE ON VIABILITY OF FORAGE SEEDS STORED IN HERMETICALLY SEALED CANS. Agron. Jour. 51: 680–684.

——— and Schoolcraft, I.
1968. PHYSIOLOGICAL AND CHEMICAL DIFFERENCES IN AGED SEEDS. Crop Sci. 8: 407–409.

——— Taylor, H. L., and Rowell, P. T.
1960. CHANGE OF FORAGE SEED QUALITY UNDER DIFFERENT SIMULATED SHIPPING CONDITIONS. Agron. Jour. 52: 37–40.

Chirkovskiy, V. I.
1953. [THE INFLUENCE OF AGING OF SEEDS ON THE DEVELOPMENT OF TOBACCO PLANTS.] Akad. Nauk. SSSR Dok. 92: 439–442. [In Russian.] (From Hort. Abs. v. 24, No. 1658. 1954.)

———
1956. [THE DEVELOPMENT OF TOBACCO PLANTS IN RELATION TO THE AGE OF SEEDS AND SOME OTHER FACTORS.] Krasnodar Vsesoyuzn. Nauch. Issled. Inst. Tabaka i Makhorki, Sborn. Nauch. Issled. Rabot No. 149: 147–161. [In Russian.] (From Hort. Abs. v. 28, No. 2728. 1958.)

Chopinet, R.
1952. LA CONSERVATION DES SEMENCES EN ATMOSPHERE SECHE. SON INTERET EN SELECTION. 13th Internatl. Hort. Cong. 1952, pp. 1022–1043.

Christensen, C. M.
1957. DETERIORATION OF STORED GRAINS BY FUNGI. Bot. Rev. 23: 108–134.

———
1969. CONDITION AND STORABILITY OF EXPORT GRAINS. Minn. Sci. 25: 20–21.

———
1970. MOISTURE CONTENT, MOISTURE TRANSFER, AND INVASION OF STORED SORGHUM SEEDS BY FUNGI. Phytopathology 60: 280–283.

——— and Gordon, D. R.
1948. THE MOLD FLORA OF STORED WHEAT AND CORN AND ITS RELATION TO HEATING OF MOIST GRAIN. Cereal Chem. 25: 40–51.

——— and Hodson, A. C.
1960. DEVELOPMENT OF GRANARY WEEVILS AND STORAGE FUNGI IN COLUMNS OF WHEAT.—II. Jour. Econ. Ent. 53: 375–380.

——— and Kaufmann, H. H.
1965. DETERIORATION OF STORED GRAINS BY FUNGI. Ann. Rev. Phytopath. 3: 69–84.

——— and Kaufmann, H. H.
1969. GRAIN STORAGE: THE ROLE OF FUNGI IN QUALITY LOSS. 153 pp., illus. Minneapolis, Minn.

——— and Kaufmann, H. H.
1974. MICROFLORA. *In* Christensen, C. M., Storage of Cereal Grains and Their Products, 2d ed., pp. 158–192. St. Paul, Minn.

Christidis, B. G.
1954. SEED VITALITY AND OTHER COTTON CHARACTERS AS AFFECTED BY THE AGE OF SEED. Plant Physiol. 29: 124–131.

——— 1955. DORMANCY IN COTTON SEED. Agron. Jour. 47: 400–403.

CLARK, B. E.
1948. NATURE AND CAUSES OF ABNORMALITIES IN ONION SEED GERMINATION. Cornell Univ. Agr. Expt. Sta. Mem. 282, 27 pp.

CLARK, C. F.
1940. LONGEVITY OF POTATO SEED. Amer. Potato Jour. 17: 147–152.

CLARK, D. C., and BASS, L. N.
1975. EFFECTS OF STORAGE CONDITIONS, PACKAGING MATERIALS, AND MOISTURE CONTENT ON LONGEVITY OF CRIMSON CLOVER SEEDS. Crop Sci. 15: 577–580.

——— BASS, L. N., and SAYERS, R. L.
1963. STORAGE OF HEMP AND KENAF SEED. Assoc. Off. Seed Anal. Proc. 53: 210–214.

CLAYTON, E. E.
1931. EFFECT OF SEED TREATMENTS ON SEED LONGEVITY. (Abstract) Phytopathology 21: 105–106.

COBB, R. D.
1958. SEED GERMINATION AFTER FUMIGATION WITH METHYL BROMIDE FOR KHAPRA BEETLE CONTROL. Calif. Dept. Agr. Quart. Bul. 47, 19 pp.

COE, H. S., and MARTIN, J. N.
1920. STRUCTURE AND CHEMICAL NATURE OF THE SEED COAT AND ITS RELATION TO IMPERMEABLE SEEDS OF SWEET CLOVER. U.S. Dept. Agr. Bul. 844, pt. II, 39 pp.

COLEMAN, F. B., and PEEL, A. C.
1952. STORAGE OF SEEDS. Queensland Agr. Jour. 74: 265–276.

COOKE, F. M.
1966. HUMIDITY CONTROL OF SEED STORAGE AREAS. Miss. Seedsmen Assoc. Short Course Seedsmen Proc. 1966, pp. 29–37.

COOPER, C. C.
1959. POLYETHYLENE PROTECTIVE SEED PACKAGES. Amer. Soc. Hort. Sci. Proc. 74: 569–579.

COTTON, R. T.
1954. INSECTS. *In* Anderson, J. A., and Alcock, A. W., eds., Storage of Cereal Grains and Their Products, pp. 221–274, illus. St. Paul, Minn.

———
1960. PESTS OF STORED GRAIN AND GRAIN PRODUCTS. 306 pp., illus. Minneapolis, Minn.

CROCIONI, A.
1950. DURATE DEL POTERE GERMINATIVO DEL SEME DI CANAPA IN RAPPORTO ALLE CONDIZIONI DI CONSERVAZIONE. [Duration of viability of hemp seed in relation to storage conditions.] Bologna Univ. Cent. di Studi Ricerche sulla Lavorazione, Colt. ed Econ. della Canapa, Quad. 9, 30 pp.

CROCKER, WILLIAM.
1938. LIFE SPAN OF SEEDS. Bot. Rev. 4: 235–274.

———
1948. GROWTH OF PLANTS. 459 pp. Rheinhold, New York.

——— and BARTON, L. V.
1953. PHYSIOLOGY OF SEEDS. 267 pp., illus. Waltham, Mass.

——— and HARRINGTON, G. T.
1918. CATALASE AND OXIDASE CONTENT OF SEEDS IN RELATION TO THEIR DORMANCY, AGE, VITALITY, AND RESPIRATION. Jour. Agr. Res. 15: 137.

CROSIER, W. F.
1938. ABNORMAL GERMINATION OF WHEAT CAUSED BY ORGANIC MERCURIALS. Assoc. Off. Seed Anal. Proc. 26 (1933): 284–285.
——— and PATRICK, S.
1952. LONGEVITY OF HARD SEEDS IN WINTER VETCH. Assoc. Off. Seed Anal. Proc. 42: 75–80.
D'AMATO, F., and HOFFMAN-OSTENHOF, O.
1956. METABOLISM AND SPONTANEOUS MUTATIONS IN PLANTS. *In* Demerec, M., ed., Adv. in Genet. 8: 1–28.
DARLINGTON, H. T.
1931. THE 50-YEAR PERIOD FOR DR. BEAL'S SEED VIABILITY EXPERIMENT. Amer. Jour. Bot. 18: 262–265.
——— and STEINBAUER, G. P.
1960. THE EIGHTY-YEAR PERIOD OF THE BEAL BURIED SEED EXPERIMENT. (Abstract) Plant Physiol. 35 (sup.): xxxi.
DAVIES, W. E.
1956. STORAGE OF CLOVER SEED. I. FIRST INTERIM REPORT 1947–54. Brit. Grassland Soc. Jour. 11: 224–229.
DAVIS, N. D.
1961. PEANUT STORAGE STUDIES. CHEMICAL CHANGES DURING 8 YEARS OF STORAGE. Ala. Acad. Sci. Jour. 32: 251–254.
DAVIS, W. C.
1931. PHENALASE ACTIVITY IN RELATION TO SEED VIABILITY. Plant Physiol. 6: 127–138.
DAVIS, W. E.
1926. THE USE OF CATALASE AS A MEANS OF DETERMINING THE VIABILITY OF SEEDS. Boyce Thompson Inst. Plant Res. Prof. Paper 1: 6–12.
DE CANDOLLE, AUG.-PYR.
1832. CONSERVATION DES GRAINES. *His* Physiologie Végétale, v. 2, pp. 618–626. Paris.
———
1846. SUR LA DUREE RELATIVE DE LA FACULTE DE GERMER DANS DES GRAINES APPARTENANT A DIVERSES FAMILLES. Ann. des Sci. Nat., Bot. III 6: 373–382.
DE CANDOLLE, C., and PICTET, R.
1895. SUR LA VIE LATENT DES GRAINES. Arch. des Sci., Phys. et Nat. 33: 497–512.
DELOUCHE, J. C., and BASKIN, C. C.
1971. DEVELOPMENT OF METHODS FOR PREDICTING THE LONGEVITY OF CROP SEED LOTS IN STORAGE. Coop. Agreement 12–14–100–9010 (51), Terminal Rpt. Reported to Mkt. Qual. Res. Div., Agr. Res. Serv., U.S. Dept. Agr., 200 pp. Miss. State Univ., State College.
——— RUSHING, T. T., and BASKIN, C. C.
1967. PREDICTING THE RELATIVE STORABILITY OF SEED LOTS. Report to the American Seed Research Foundation, 26 pp. Miss. State Univ., State College.
——— RUSHING, T. T., and BASKIN, C. C.
1968. PREDICTING THE RELATIVE STORABILITY OF CROP SEED LOTS. Search 8 (1): 1, 6–9.
DE TEMPE, J.
1961. ROUTINE METHODS FOR DETERMINING THE HEALTH CONDITION OF SEEDS IN THE SEED TESTING STATION. Internatl. Seed Testing Assoc. Proc. 26: 27–60.

De Witt, J. L., Canode, C. L., and Patterson, J. K.
1962. EFFECTS OF HEATING AND STORAGE ON THE VIABILITY OF GRASS SEED HARVESTED WITH HIGH MOISTURE CONTENT. Agron. Jour. 54: 126–129.

Dexter, S. T.
1957. MOISTURE EQUILIBRIUM VALUES IN RELATION TO MOLD FORMATION OF SEEDS OF SEVERAL GRASS AND SMALL-SEEDED LEGUMES. Agron. Jour. 49: 485–488.

——— and Creighton, J. W.
1948. A METHOD OF CURING FARM PRODUCTS BY THE USE OF DRYING AGENTS. Amer. Soc. Agron. Jour. 40: 70–79.

Diener, U. L.
1958. THE MYCOFLORA OF STORED PEANUTS. Ala. Acad. Sci. Jour. 30 (2): 5–6.

———
1960. THE MYCOFLORA OF PEANUTS IN STORAGE. Phytopathology 50: 220–223.

Dillman, A. C.
1930. HYGROSCOPIC MOISTURE OF FLAX SEED AND WHEAT AND ITS RELATION TO COMBINE HARVESTING. Amer. Soc. Agron. Jour. 22: 51–74.

——— and Toole, E. H.
1937. EFFECT OF AGE, CONDITION, AND TEMPERATURE ON THE GERMINATION OF FLAXSEED. Amer. Soc. Agron. Jour. 29: 23–29.

Dimmock, F.
1947. THE EFFECTS OF IMMATURITY AND ARTIFICIAL DRYING UPON THE QUALITY OF SEED CORN. Canad. Dept. Agr. Tech. Bul. 58, 66 pp.

Dolan, D. D., Braverman, S. W., and Pfleger, F. L.
1973. A WALK-IN DRYING ROOM FOR REMOVING MOISTURE FROM SEED CROPS OF FORAGE INTRODUCTIONS. Agron. Jour. 63: 678–680.

Dore, J.
1955. DORMANCY AND VIABILITY OF PADI SEED. Malay Agr. Jour. 38: 163–173.

Dorph-Petersen, K.
1928. COMBIEN DE TEMPS LES SEMENCES DE TUSSILAGO FARFARA GARDENT-ELLES LEUR FACULTE GERMINATIVE SOUS DE DIFFERENTES CONDITIONS DE TEMPERATURE? Internatl. Seed Testing Assoc. Proc. 1: 72–76.

Dorrell, D. G., and Adams, M. W.
1969. EFFECT OF SOME SEED CHARACTERISTICS ON MECHANICALLY INDUCED SEEDCOAT DAMAGE IN NAVY BEANS (PHASEOLUS VULGARIS L.). Agron. Jour. 61: 672–673.

Dossat, R. J.
1961. PRINCIPLES OF REFRIGERATION. 544 pp., illus. New York and London.

Dryomatic Division, LogEtronics.
1965. DEHUMIDIFICATION ENGINEERING MANUAL. 72 pp. Gaithersburg, Md.

Dutt, B. K., and Thakurta, A. G.
1956. VIABILITY OF VEGETABLE SEEDS IN STORAGE. Bose Res. Inst., Calcutta, Trans. 19 (1953–55): 27–36.

Duvel, J. W. T.
1905. THE VITALITY OF BURIED SEEDS. U.S. Dept. Agr. Bur. Plant Indus. Bul. 83, 20 pp.

Edmond, J. B.
1959. THE INFLUENCE OF CERTAIN STORAGE CONDITIONS AND CALCIUM OXIDE ON THE STORAGE LIFE OF GOLDEN CROSS BANTAM SWEET CORN SEED. Amer. Hort. Sci. Proc. 74: 415–421.

Eguchi, T., and Yamada, H.
1958. [STUDIES ON THE EFFECT OF MATURITY ON LONGEVITY IN VEGETABLE SEEDS.] Natl. Inst. Agr. Sci. Bul., ser. E, Hort. 7: 145–165. [In Japanese. English summary.]

ESBO, H.

1954. LIVSKRAFTENS BIBEHALLANDE HOS OSKALAT OCH SKALAT FRO AV TIMOTJ VID LANGTIDSLAGRING UNDER ORDINARA BETINGELSER PA FROMAGASIN. K. Landtbr. Akad. Tidskr. 93: 123–148. [English summary, pp. 146–148.]

——— 1959. LIVSKRAFTEN HOS TIMOTEJFRO UNDER LANGTIDSLAGRING. Lantbrhogsk. Inst. f. Växtodlingslara No. 12, 101 pp.

ESDORN, I., and STUETZ, H.

1932. DIE BEWERTUNG HARTER LEGUMINOSENSAMEN. Landw. Vers. Sta. 114: 137–147.

EVANS, G.

1957a. RED CLOVER SEED STORAGE FOR 23 YEARS. Brit. Grassland Soc. Jour. 12: 171–177.

——— 1957b. THE VIABILITY OVER A PERIOD OF FIFTEEN YEARS OF SEVERELY DRIED RYEGRASS SEED. Brit. Grassland Soc. Jour. 12: 286–289.

——— ROBERTS, H. M., and LEWIS, J.

1958. MAINTENANCE OF LIFE IN SEEDS. Welsh Plant Breeding Sta., Aberystwyth Rpt. 1950–56: 152–155.

EWART, A. J.

1908. ON THE LONGEVITY OF SEEDS. Roy. Soc. Victoria, Proc. 21: 2–210.

FIELDS, R. W., and KING, T. H.

1962. INFLUENCE OF STORAGE FUNGI ON DETERIORATION OF STORED PEA SEED. Phytopathology 52: 336–339.

FIFIELD, C. C., and ROBERTSON, D. W.

1952. MILLING, BAKING, AND CHEMICAL PROPERTIES OF MARQUIS AND KANRED WHEAT GROWN IN COLORADO AND STORED 19 TO 27 YEARS. Agron. Jour. 44: 555–559.

FILTER, P.

1932–33. UNTERSUCHUNGEN UBER DIE LEBENSDAUER VON HANDELSUND ANDEREN SAATEN, MIT BESONDERER BERUCKSICHTIGUNG DER HARTEN SAMEN UND DER ALTERSVERFARBUNG BEI DEN LEGUMINOSEN. Landw. Vers. Sta. 114: 149–170.

FISHER, E. A., and JONES, C. R.

1939. A NOTE ON MOISTURE INTERCHANGE IN MIXED WHEATS, WITH OBSERVATIONS ON THE RATE OF ABSORPTION OF MOISTURE BY WHEAT. Cereal Chem. 16: 573–583.

FLORIS, C.

1970. AGEING IN TRITICUM DURUM SEEDS: BEHAVIOR OF EMBRYOS AND ENDOSPERMS FROM AGED SEEDS AS REVEALED BY THE EMBRYO-TRANSPLANTATION TECHNIQUE. Jour. Expt. Bot. 21: 462–468.

FORNEROD, W. P.

1963. TESTING OF PACKAGES AND PACKAGING MATERIALS FOR SOWING SEEDS ON THEIR BEHAVIOUR TO WATER VAPOUR PERMEABILITY. Internatl. Seed Testing Assoc. Proc. 28: 935–942.

FOY, N. R.

1934. DETERIORATION PROBLEMS IN NEW ZEALAND CHEWINGS FESCUE. New Zeal. Jour. Agr. 49: 10–24.

FREY, E.

1960. DER WASSERGEHALT UND DIE KEIMFAHIGKEIT VON WEIZEN UNTER VERSCHIEDEN LAGERUNGSBEDINGUNGEN UND AN VERSCHIEDENEN STELLEN INNERHALB DER SACKE. Internatl. Seed Testing Assoc. Proc. 25: 555–566.

FULUTOWICZ, A., and BEJNAR, W.
1954. [THE INFLUENCE OF THE AGE OF SUGAR BEET SEEDS ON THEIR SEEDING VALUE.] Rocz. Nauk. Rolnicz., Ser. A, 69: 323–340. [In Polish.]

GAERTNER, E. E.
1950. STUDIES OF SEED GERMINATION, SEED IDENTIFICATION, AND HOST RELATIONSHIPS IN DODDERS, CUSCUTA SPP. N.Y. (Cornell) Agr. Expt. Sta. Mem. 294, 56 pp.

GANE, R.
1941. THE WATER CONTENT OF WHEATS AS A FUNCTION OF TEMPERATURE AND HUMIDITY. Soc. Chem. Indus. Jour., Trans. and Commun. 60: 44–46.

———
1948a. THE EFFECT OF TEMPERATURE, HUMIDITY AND ATMOSPHERE ON THE VIABILITY OF CHEWING'S FESCUE GRASS SEED IN STORAGE. Jour. Agr. Sci. 38: 90–92.

———
1948b. THE EFFECT OF TEMPERATURE, WATER CONTENT AND COMPOSITION OF THE ATMOSPHERE ON THE VIABILITY OF CARROT, ONION AND PARSNIP SEEDS IN STORAGE. Jour. Agr. Sci. 38: 84–89.

GARREN, K. H.
1966. PEANUT (GROUNDNUT) MICROFLORAS AND PATHOGENESIS IN PEANUT POD ROT. Phytopath. Ztschr. 55: 359–367.

GAY, F. J.
1946. THE EFFECT OF TEMPERATURE ON THE MOISTURE CONTENT-RELATIVE HUMIDITY EQUILIBRIA OF WHEAT. Austral. Council Sci. and Indus. Res. Jour. 19: 187–189.

GHOSH, T., and BASAK, M.
1958. METHOD OF STORING JUTE SEED AND EFFECT OF AGE OF SEED ON YIELD OF FIBRE. Indian Jour. Agr. Sci. 38: 235–242.

——— BASAK, M., and KUNDU, B. C.
1951. EFFECT OF INSULATION AND CHEMICAL TREATMENT IN RELATION TO STORING OF JUTE SEEDS. Indian Phytopath. 4: 38–44.

GILL, N. S., and DELOUCHE, J. C.
1973. DETERIORATION OF SEED CORN DURING STORAGE. (Abstract) Assoc. Off. Seed Anal. News Letter 47 (2): 23.

GILMAN, J. C., and BARRON, D. H.
1930. EFFECT OF MOLDS ON TEMPERATURE OF STORED GRAIN. Plant Physiol. 5: 565–573.

GLASS, R. L., PONTE, J. G., JR., CHRISTENSEN, C. M., and GEDDES, W. F.
1959. GRAIN STORAGE STUDIES. XXVIII. THE INFLUENCE OF TEMPERATURE AND MOISTURE LEVEL ON THE BEHAVIOR OF WHEAT STORED IN AIR OR NITROGEN. Cereal Chem. 36: 341–356.

GODWIN, H., and WILLIS, E. H.
1964. THE VIABILITY OF LOTUS SEEDS (NELUMBIUM NUCIFERA, GAERTN.). New Phytol. 63: 410–412.

GOFF, E. S.
1890. COMPARATIVE VITALITY OF HULLED AND UNHULLED TIMOTHY SEEDS. Wis. Agr. Expt. Sta. Ann. Rpt. 7: 202–204.

GOFF, J.
1972. ACCELERATED AGING: DESIGN TECHNIQUE AND PRACTICAL APPLICATIONS. Search 11: 5–8.

GOODSELL, S. F., HUEY, G., and ROYCE, R.
1955. THE EFFECT OF MOISTURE AND TEMPERATURE DURING STORAGE ON COLD TEST REACTION OF ZEA MAYS SEED STORED IN AIR, CARBON DIOXIDE, OR NITROGEN. Agron. Jour. 47: 61–64.

Goss, W. L.
1926. GERMINATION OF HARD SEEDS OF HAIRY VETCH. Assoc. Off. Seed Anal. Proc. 18: 42–43.

——— 1937. GERMINATION OF FLOWER SEEDS STORED FOR TEN YEARS IN THE CALIFORNIA STATE SEED LABORATORY. Calif. Dept. Agr. Bul. 26, pp. 326–333.

GRABE, D. F.
1957. ARTIFICIAL DRYING AND STORAGE OF SMOOTH BROMEGRASS SEED. Agron. Jour. 49: 161–165.

——— 1964. GLUTAMIC ACID DECARBOXYLASE ACTIVITY AS A MEASURE OF SEEDLING VIGOR. Assoc. Off. Seed Anal. Proc. 54: 100–109.

——— 1970. TETRAZOLIUM TESTING HANDBOOK. CONTRIB. NO. 29 TO HANDBOOK ON SEED TESTING. 62 pp. Assoc. Off. Seed Anal., Corvallis, Oreg.

——— and ISELY, DUANE.
1969. SEED STORAGE IN MOISTURE-RESISTANT PACKAGES. Seed World 104 (2): 2, 4, 5.

GRABER, L. F.
1922. SCARIFICATION AS IT AFFECTS LONGEVITY OF ALFALFA SEED. Amer. Soc. Agron. Jour. 14: 298–302.

GREEN, J. G.
1961. HIGH TEMPERATURE STORAGE OF DRY MAIZE. Iowa State Univ. Diss. Abs. 22: 3358–3359.

GRIFFITH, W. L., and HARRISON, C. M.
1954. MATURITY AND CURING TEMPERATURES AND THEIR INFLUENCE ON GERMINATION OF REED CANARY GRASS SEED. Agron. Jour. 46: 163–167.

GROVES, J. F.
1917. TEMPERATURE AND LIFE DURATION OF SEEDS. Bot. Gaz. 63: 169–189.

GUILLAUMIN, A.
1928. LE MAINTIEN DES GRAINES DANS UN MILIEU PRIVE D'OXYGENE COMME MOYEN DE PROLONGER LEUR FACULTE GERMINATIVE. [Paris] Acad. des Sci. Compt. Rend. 187: 571–572.

GUNTHARDT, H., SMITH, L., HAFERKAMP, M. E., and NILAN, R. A.
1953. STUDIES ON AGED SEEDS. II. RELATION OF AGE OF SEEDS TO CYTOGENIC EFFECTS. Agron. Jour. 45: 438–441.

HABER, E. S.
1950. LONGEVITY OF THE SEED OF SWEETCORN INBREDS AND HYBRIDS. Amer. Soc. Hort. Sci. Proc. 55: 410–412.

HAFERKAMP, M. E., SMITH, LUTHER, and NILAN, R. A.
1953. STUDIES ON AGED SEEDS. I. RELATION OF AGE OF SEED TO GERMINATION AND LONGEVITY. Agron. Jour. 45: 434–437, illus.

HALL, C. W.
1957. DRYING FARM CROPS. 336 pp., illus. Reynoldsburg, Ohio.

HANLIN, R. T.
1966. CURRENT RESEARCH ON PEANUT FUNGI IN GEORGIA. Ga. Agr. Res. 8 (2): 3–4.

HANSON, C. H., and MOORE, R. P.
1959. VIABILITY OF SEEDS OF EIGHT FORAGE CROP PLANTS STORED UNDER SUBFREEZING CONDITIONS. Agron. Jour. 51: 627.

HARRINGTON, G. T.
1915. HARD CLOVER SEED AND ITS TREATMENT IN HULLING. U.S. Dept. Agr. Farmers' Bul. 676, 8 pp.

———
1916. AGRICULTURAL VALUE OF IMPERMEABLE SEED. Jour. Agr. Res. 6: 761–796.

——— and CROCKER, W.
1918. RESISTANCE OF SEEDS TO DESICCATION. Jour. Agr. Res. 14: 525–532.

HARRINGTON, J. F.
1949. HARD SEED IN BEANS AND OTHER LEGUMES. Seed World 64: 42, 44.

———
1960a. PRELIMINARY REPORT ON THE RELATIVE DESIRABILITY OF DIFFERENT CONTAINERS FOR STORAGE OF SEVERAL KINDS OF VEGETABLE SEED. Calif. Univ. Dept. Veg. Crops Ser. 104, 8 pp.

———
1960b. DRYING, STORING, AND PACKAGING SEEDS TO MAINTAIN GERMINATION AND VIGOR. Miss. State Univ. Short Course Seedsmen Proc. 1959, pp. 89–107.

———
1960c. DRYING, STORING AND PACKAGING SEED TO MAINTAIN GERMINATION. Seedsmen's Digest 1: 16, 56–57, 64, 66, 68.

———
1960d. THUMB RULES OF DRYING SEED. Crops and Soils 13: 16–17.

———
1963. THE VALUE OF MOISTURE-RESISTANT CONTAINERS IN VEGETABLE SEED PACKAGING. Calif. Agr. Expt. Sta. Bul. 792, 23 pp.

———
1967. SEED AND POLLEN STORAGE FOR CONSERVATION OF PLANT GENE RESOURCES. FAO Tech. Conf. Exploration, Utilization, Conservation Plant Gene Resources, Sect. 13 (iii), pp. 1–22.

———
1968. MOISTURE EQUILIBRIUM VALUES FOR SEVERAL GRASS AND LEGUME SEEDS. Agron. Jour. 60: 594–597.

———
1972. SEED STORAGE AND LONGEVITY. *In* Kozlowski, T. T., Seed Biology, v. 3, pp. 145–245, illus. New York and London.

———
1973. PROBLEMS OF SEED STORAGE. *In* Heydecker, W., Seed Ecology, pp. 251–263. Pa. State Univ. Press, University Park, Pa., and London.

——— and AGUIRRE, M.
1963. SPEED OF MOISTURE PENETRATION INTO VEGETABLE SEED. Seed World 92 (8): 8–9.

——— and DOUGLAS, J. E.
1970. SEED STORAGE AND PACKAGING—APPLICATIONS FOR INDIA. 222 pp. New Delhi, India.

——— and THOMPSON, R. C.
1952. EFFECT OF VARIETY AND AREA OF PRODUCTION ON SUBSEQUENT GERMINATION OF LETTUCE SEED AT HIGH TEMPERATURES. Amer. Soc. Hort. Sci. Proc. 59: 445–450.

HARRISON, B. J.
1956. SEED STORAGE. John Innes Hort. Inst. Ann. Rpt. 46: 15–16.

——— and MCLEISH, J.
1954. ABNORMALITIES OF STORED SEED. Nature 173: 593–594.

HART, J. R., FEINSTEIN, L., and GOLUMBIC, C.
1959. OVEN METHODS FOR PRECISE MEASUREMENT OF MOISTURE IN SEEDS. U.S. Dept. Agr. Mktg. Res. Rpt. 304, 16 pp.

HARTER, L. L.

1930. THRESHER INJURY A CAUSE OF BALDHEAD IN BEANS. Jour. Agr. Res. 40: 371–384.

HASS, A.

1961. DEHUMIDIFICATION IN SEED STORAGE. Seedsmen's Digest 12 (9): 8, 10, 11, 32, 33.

———

1965. HUMIDITY CONTROL IN SEED STORAGE. Miss. State Univ. Short Course Seedsmen Proc. 1965, pp. 157–160.

HAYNES, B. C., JR.

1961. VAPOR PRESSURE DETERMINATION OF SEED HYGROSCOPICITY. U.S. Dept. Agr. Tech. Bul. 1229, 20 pp.

HEINRICH, M.

1913. DER EINFLUSS DER LUFTFEUCHTIGKEIT, DER WARME UND DER SAUERSTOFFS DER LUFT AUF LAGERNDES. Saatgut Landw. Vers. Sta. 81: 289–376.

HEIT, C. E., and MUNN, M. T.

1947. EFFECT OF LOW SEED VITALITY ON YIELD. Farm Res. 13 (2): 14–15.

HELGESON, E. A.

1932. IMPERMEABILITY IN MATURE AND IMMATURE SWEET CLOVER SEEDS AS AFFECTED BY CONDITIONS OF STORAGE. Wis. Acad. Sci. Trans., Arts and Letters 27: 193–206.

HENDERSON, L. S., and CHRISTENSEN, C. M.

1961. POSTHARVEST CONTROL OF INSECTS AND FUNGI. U.S. Dept. Agr. Ybk. 1961: 348–356.

HENDERSON, S. M.

1952. A BASIC CONCEPT OF EQUILIBRIUM MOISTURE. Agr. Engin. 33: 29–32.

HERMANN, E. M., and HERMANN, W.

1939. THE EFFECT OF MATURITY AT TIME OF HARVEST ON CERTAIN RESPONSES OF SEED OF CRESTED WHEATGRASS, AGROPYRON CRISTATUM (L.) GAERTN. Amer. Soc. Agron. Jour. 31: 876–885.

HEYDECKER, W.

1973. SEED ECOLOGY. 578 pp. Pa. State Univ. Press, University Park, Pa., and London.

HILL, A. W.

1917. THE FLORA OF THE SOMME BATTLEFIELD. Kew Bul. 1917: 297–300.

HLYNKA, I., and ROBINSON, A. D.

1954. MOISTURE AND ITS MEASUREMENT. *In* Anderson, J. A., and Alcock, A. W., eds., Storage of Cereal Grains and Their Products, pp. 1–45, illus. St. Paul, Minn.

HOGAN, J. T., and KARON, M. L.

1955. HYGROSCOPIC EQUILIBRIUM OF ROUGH RICE AT ELEVATED TEMPERATURES. Agr. and Food Chem. 3: 855–860.

HOLMAN, L. E., and CARTER, D. G.

1952. SOYBEAN STORAGE IN FARM-TYPE BINS: A RESEARCH REPORT. Ill. Agr. Expt. Sta. Bul. 553, pp. 449–496.

HOLMES, A. D.

1953. GERMINATION OF SEEDS REMOVED FROM MATURE AND IMMATURE BUTTERNUT SQUASHES AFTER SEVEN MONTHS OF STORAGE. Amer. Soc. Hort. Sci. Proc. 62: 433–436.

HUBBARD, J. E., EARLE, F. R., and SENTI, F. R.

1957. MOISTURE RELATIONS IN WHEAT AND CORN. Cereal Chem. 34: 422–433.

HULL, F. H.
1937. INHERITANCE OF REST PERIOD OF SEEDS AND CERTAIN OTHER CHARACTERS IN THE PEANUT. Fla. Agr. Expt. Sta. Tech. Bul. 314, 46 pp.

HURD, A. M.
1921. SEED-COAT INJURY AND VIABILITY OF SEEDS OF WHEAT AND BARLEY AS FACTORS IN SUSCEPTIBILITY TO MOLDS AND FUNGICIDES. Jour. Agr. Res. 21: 99–122.

HYDE, E. O. C.
1954. THE FUNCTION OF THE HILUM IN SOME PAPILIONACEAE IN RELATION TO THE RIPENING OF THE SEED AND THE PERMEABILITY OF THE TESTA. Ann. Bot. (N.S.) 18: 241–256.

INOUE, Y., and SUZUKI, Y.
1962. STUDIES ON THE EFFECTS OF MATURITY AND AFTER-RIPENING OF SEEDS UPON THE SEED GERMINATION IN SNAP BEANS, PHASEOLUS VULGARIS L. Jap. Soc. Hort. Sci. Jour. 31: 146–150.

INTERNATIONAL SEED TESTING ASSOCIATION.
1966. INTERNATIONAL RULES FOR SEED TESTING. Internatl. Seed Testing Assoc. Proc. 31: 1–152. (For amendments to 1966 rules, see ibid. 37: 187–209 (1972).)

ISELY, DUANE, and BASS, L. N.
1960. SEEDS AND PACKAGING MATERIALS. 14th Hybrid Corn Indus. Res. Conf. Proc. 1959, pp. 101–110.

ITO, HIROSHI.
1972. ORGANISATION OF THE NATIONAL SEED STORAGE LABORATORY FOR GENETIC RESOURCES IN JAPAN. *In* Roberts, E. H., Viability of Seeds, app. 2, pp. 405–416, illus. London.

——— and HAYASHI, KEN-ICHI.
1960. STUDIES ON THE STORAGE OF RICE SEEDS. I. INFLUENCES OF TEMPERATURE AND MOISTURE CONTENT ON THE LONGEVITY OF RICE SEEDS AND METHODS OF DRYING. Crop Sci. Soc. Japan, Proc. 28: 363–364. [In Japanese. English summary.]

JACKSON, C. R.
1965. PEANUT-POD MYCOFLORA AND KERNEL INFECTION. Plant and Soil 23: 203–212.

JAMES, EDWIN.
1962. MECHANICAL CONTROL OF HUMIDITY IN SEED STORAGE ROOMS. Seedsmen's Digest (Dec.), pp. 12, 14, 40.

———
1967. PRESERVATION OF SEED STOCKS. Adv. in Agron. 19: 87–106.

———
1968. LIMITATIONS OF GLUTAMIC ACID DECARBOXYLASE ACTIVITY FOR ESTIMATING VIABILITY IN BEANS (PHASEOLUS VULGARIS L.). Crop Sci. 8: 403–404.

———
1972. ORGANISATION OF THE UNITED STATES NATIONAL SEED STORAGE LABORATORY. *In* Roberts, E. H., Viability of Seeds, app. 1, pp. 397–404, illus. London.

——— BASS, L. N., and CLARK, D. C.
1964. LONGEVITY OF VEGETABLE SEEDS STORED 15 TO 30 YEARS AT CHEYENNE, WYOMING. Amer. Soc. Hort. Sci. Proc. 84: 527–534.

——— BASS, L. N., and CLARK, D. C.
1967. VARIETAL DIFFERENCES IN LONGEVITY OF VEGETABLE SEEDS AND THEIR RESPONSE TO VARIOUS STORAGE CONDITIONS. Amer. Soc. Hort. Sci. Proc. 91: 521–528.

JENSEN, CARLOS.
1941. TREATING SEED WITH LIGHT. Seed World 49 (9): 20–21.

———
1942. UBER DIE MOGLICHKEIT, MIT HILFE VON LICHTBEHANDLUNG, DIE KEIMFAHIGKEIT VON SAMEN ZU VERLANGERN. Ztschr. f. Bot. 37: 487–499.

JENSEN, H. A., and JORGENSEN, J.
1969. THE INFLUENCE OF THE DEGREE OF MATURITY AND DRYING ON THE GERMINATING CAPACITY OF FESTUCA PRATENSIS HUDS. Acta Agr. Scand. 19: 258–264.

JOSEPH, H. C.
1929. GERMINATION AND KEEPING QUALITY OF PARSNIP SEEDS UNDER VARIOUS CONDITIONS. Bot. Gaz. 87: 195–210. (Also Boyce Thompson Inst. Contrib. 2: 115–130.)

JOUBERT, T. G. LA G.
1954. HARD-SKIN IN BEANS. Farming in So. Africa 29: 225, 232.

JUSTICE, O. L.
1950. THE TESTING FOR PURITY AND GERMINATION OF SEED OFFERED FOR IMPORTATION INTO THE UNITED STATES. Internatl. Seed Testing Assoc. Proc. 16: 156–172.

———
1956. GERMINATION BEHAVIOR IN SEEDS OF NUTGRASS (CYPERUS ROTUNDUS L.). Assoc. Off. Seed Anal. Proc. 46: 67–71.

———
1972. ESSENTIALS OF SEED TESTING. *In* Kozlowski, T. T., Seed Biology, v. 3, pp. 301–370, illus. New York and London.

——— and WHITEHEAD, M. D.
1942. VIABILITY OF VELVETBEAN, STIZOLOGIUM SPP., SEED AS AFFECTED BY DATE OF HARVEST, WEATHERING, STORAGE, AND LODGING. Amer. Soc. Agron. Jour. 34: 1000–1009.

KAIHARA, HIROMICHI.
1951. COMPARATIVE STORAGE-TOLERANCE OF SOME CEREAL GRAINS IN JAPAN. Ohara Inst. f. Landw. Forsch. Ber. 9: 435–440.

KALOYEREAS, S. A.
1955. A GAS STORAGE FOR PRESERVING ROUGH RICE. Rice Jour. 58: 40–42.

KANTOR, D. J., and WEBSTER, O. J.
1967. EFFECTS OF FREEZING AND MECHANICAL INJURY ON VIABILITY OF SORGHUM SEED. Crop Sci. 7: 196–199.

KARON, M. L., and HILLERY, B. E.
1949. HYGROSCOPIC EQUILIBRIUM OF PEANUTS. Amer. Oil Chem. Soc. Jour. 26: 16–19.

KARPER, R. E., and JONES, D. L.
1936. LONGEVITY AND VIABILITY OF SORGHUM SEED. Amer. Soc. Agron. Jour. 28: 330–331.

KEARNS, VIVIAN, and TOOLE, E. H.
1939. RELATION OF TEMPERATURE AND MOISTURE CONTENT TO LONGEVITY OF CHEWINGS FESCUE SEED. U.S. Dept. Agr. Tech. Bul. 670, 27 pp.

KELLY, C. F., STAHL, B. M., SALMON, S. C., and BLACK, R. H.
1942. WHEAT STORAGE IN EXPERIMENTAL FARM-TYPE BINS. U.S. Dept. Agr. Cir. 637, 245 pp.

KIESSELBACH, T. A., and RATCLIFF, J. A.
1918. FREEZING INJURY OF SEED CORN. Nebr. Agr. Expt. Sta. Bul. 163, 16 pp.

______ and RATCLIFF, J. A.

1920. FREEZING INJURY OF SEED CORN. Nebr. Agr. Expt. Sta. Res. Bul. 16, 96 pp.

KIKUCHI, M.

1954. [STUDIES ON THE DECREPITUDE OF SEEDS BY USE OF EMBRYO TRANSPLANTING METHOD. I: ON THE OLD SEEDS OF WHEAT.] Utsunomiya Univ. Col. Agr. Bul. 2: 277–292. [In Japanese. English summary.]

KINCAID, R. R.

1943. THE EFFECTS OF STORAGE CONDITIONS ON THE VIABILITY OF TOBACCO SEED. Jour. Agr. Res. 67: 407–410.

______

1958. TOBACCO SEED STORAGE FOR 25 YEARS. Fla. Agr. Expt. Sta. Tech. Bul. 593, 10 pp.

KING, D. R., GARNER, C. F., METZER, R. B., and COFFEY, L. C.

1960. EFFECT OF FUMIGATION FOR INSECT CONTROL ON SEED GERMINATION. Tex. Agr. Expt. Sta. Bul. MP–448, 12 pp.

KIVILAAN, A., and BANDURSKI, R. S.

1973. THE NINETY-YEAR PERIOD FOR DR. BEAL'S SEED VIABILITY EXPERIMENT. Amer. Jour. Bot. 60: 140–145.

KNOWLES, R. P.

1967. SUBFREEZING STORAGE OF GRASS SEEDS. Agron. Jour. 51: 86.

KOEHLER, B.

1938. EFFECT OF PROLONGED STORAGE OF TREATED SEED CORN. (Abstract) Phytopathology 28: 13.

______ and BEVER, W. M.

1954. OAT AND WHEAT SEED TREATMENTS: METHODS, INJURIES, AND GAINS. U.S. Agr. Res. Serv. Plant Dis. Rptr. 38: 762–768.

______ and BEVER, W. M.

1956. EFFECT OF FUNGICIDE AND OF STORAGE TEMPERATURE ON FUNGICIDE INJURY TO WHEAT SEED. U.S. Agr. Res. Serv. Plant Dis. Rptr. 40: 490–492.

KOLLER, DOV.

1972. ENVIRONMENTAL CONTROL OF SEED GERMINATION. *In* Kozlowski, T. T., Seed Biology, v. 2, pp. 1–101, illus. New York and London.

KONDO, MANTARO, and ISSHIKI, S.

1936. STORAGE OF RICE XIV. REMOVAL OF MOISTURE FROM THE AIR IN A GRANARY AND THE HULLED RICE STORED THEREIN BY A DESICCATING MATERIAL. Ohara Inst. f. Landw. Forsch. Ber. 7: 227–237.

______ and KASAHARA, Y.

1940. [STORAGE LIFE OF SEEDS.] Crop Sci. Soc. Japan, Proc. 12: 21–24. [In Japanese. English summary.] (Also, Jap. Jour. Bot. 11 (1941): 197.)

______ and KASAHARA, Y.

1941. [A REPORT ON STORAGE OF SEEDS FOR 10 YEARS WITH CALCIUM CHLORIDE.] Ohara Inst. Agr. Res. Spec. Rpt. 32: 304–310. [In Japanese.]

______ MATSURHIMA, S., and OKAMURA, T.

1929. KEIMKRAFT NAHRSTOFFE UND B-BITAMIN DER IM KOHLENSAUREN UND LUFTDICHTEN VERSCHLUSSE 4 JAHRE LANG AUFBEVVAHRTEN REISKORNER. Imp. Acad. Japan, Proc. 5: 159–160.

______ and OKAMURA, T.

1927. EINWIRKUNG DES KOHLENSAURE UND LUFTDICHTEN VERSCHULUS AUF DIE REISKORNER WAHREND DER AUFBEWAHRUNGSZEIT. Imp. Acad. Japan, Proc. 3: 706–709.

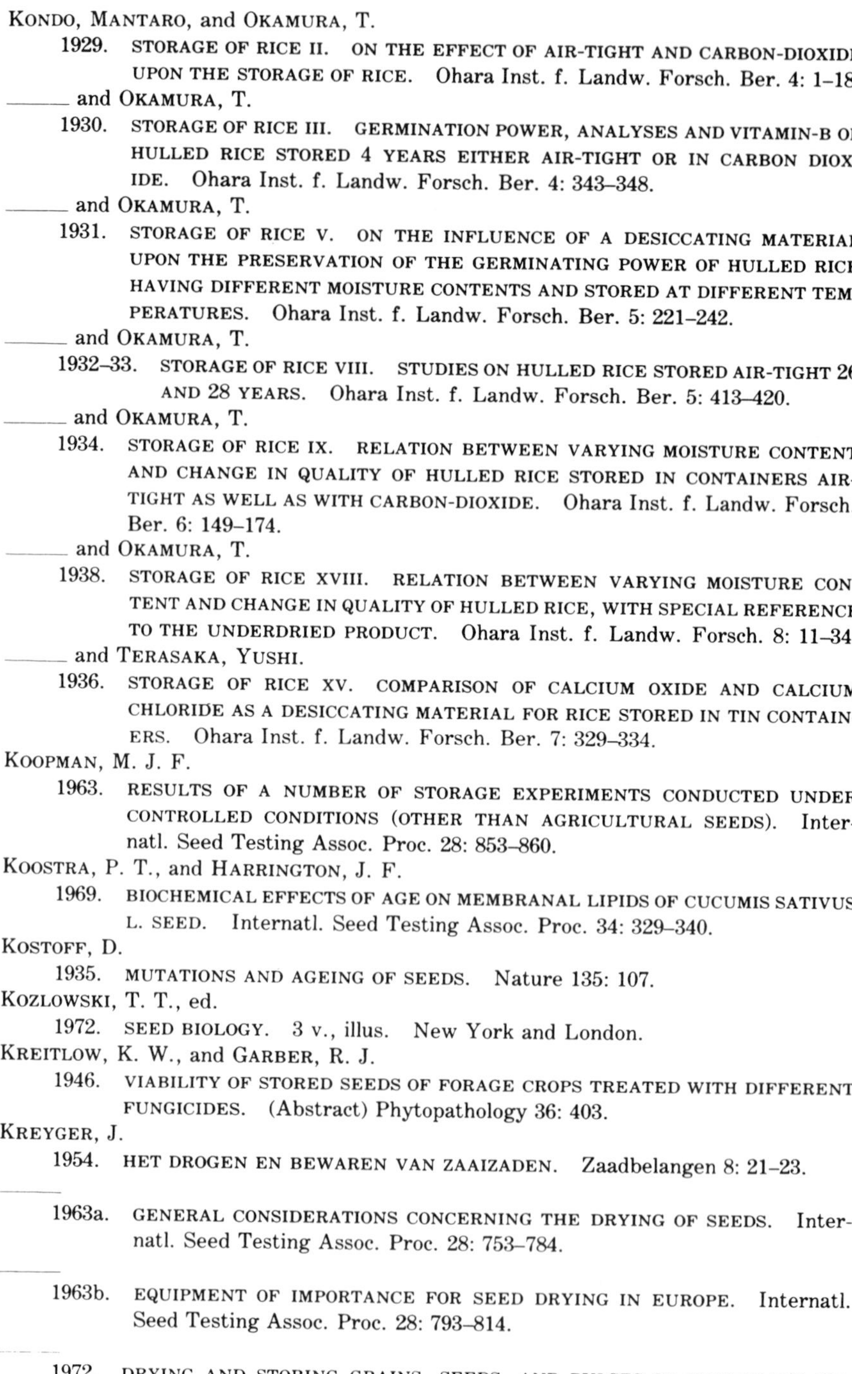

KONDO, MANTARO, and OKAMURA, T.

1929. STORAGE OF RICE II. ON THE EFFECT OF AIR-TIGHT AND CARBON-DIOXIDE UPON THE STORAGE OF RICE. Ohara Inst. f. Landw. Forsch. Ber. 4: 1–18.

——— and OKAMURA, T.

1930. STORAGE OF RICE III. GERMINATION POWER, ANALYSES AND VITAMIN-B OF HULLED RICE STORED 4 YEARS EITHER AIR-TIGHT OR IN CARBON DIOXIDE. Ohara Inst. f. Landw. Forsch. Ber. 4: 343–348.

——— and OKAMURA, T.

1931. STORAGE OF RICE V. ON THE INFLUENCE OF A DESICCATING MATERIAL UPON THE PRESERVATION OF THE GERMINATING POWER OF HULLED RICE HAVING DIFFERENT MOISTURE CONTENTS AND STORED AT DIFFERENT TEMPERATURES. Ohara Inst. f. Landw. Forsch. Ber. 5: 221–242.

——— and OKAMURA, T.

1932–33. STORAGE OF RICE VIII. STUDIES ON HULLED RICE STORED AIR-TIGHT 26 AND 28 YEARS. Ohara Inst. f. Landw. Forsch. Ber. 5: 413–420.

——— and OKAMURA, T.

1934. STORAGE OF RICE IX. RELATION BETWEEN VARYING MOISTURE CONTENT AND CHANGE IN QUALITY OF HULLED RICE STORED IN CONTAINERS AIR-TIGHT AS WELL AS WITH CARBON-DIOXIDE. Ohara Inst. f. Landw. Forsch. Ber. 6: 149–174.

——— and OKAMURA, T.

1938. STORAGE OF RICE XVIII. RELATION BETWEEN VARYING MOISTURE CONTENT AND CHANGE IN QUALITY OF HULLED RICE, WITH SPECIAL REFERENCE TO THE UNDERDRIED PRODUCT. Ohara Inst. f. Landw. Forsch. 8: 11–34.

——— and TERASAKA, YUSHI.

1936. STORAGE OF RICE XV. COMPARISON OF CALCIUM OXIDE AND CALCIUM CHLORIDE AS A DESICCATING MATERIAL FOR RICE STORED IN TIN CONTAINERS. Ohara Inst. f. Landw. Forsch. Ber. 7: 329–334.

KOOPMAN, M. J. F.

1963. RESULTS OF A NUMBER OF STORAGE EXPERIMENTS CONDUCTED UNDER CONTROLLED CONDITIONS (OTHER THAN AGRICULTURAL SEEDS). Internatl. Seed Testing Assoc. Proc. 28: 853–860.

KOOSTRA, P. T., and HARRINGTON, J. F.

1969. BIOCHEMICAL EFFECTS OF AGE ON MEMBRANAL LIPIDS OF CUCUMIS SATIVUS L. SEED. Internatl. Seed Testing Assoc. Proc. 34: 329–340.

KOSTOFF, D.

1935. MUTATIONS AND AGEING OF SEEDS. Nature 135: 107.

KOZLOWSKI, T. T., ed.

1972. SEED BIOLOGY. 3 v., illus. New York and London.

KREITLOW, K. W., and GARBER, R. J.

1946. VIABILITY OF STORED SEEDS OF FORAGE CROPS TREATED WITH DIFFERENT FUNGICIDES. (Abstract) Phytopathology 36: 403.

KREYGER, J.

1954. HET DROGEN EN BEWAREN VAN ZAAIZADEN. Zaadbelangen 8: 21–23.

———

1963a. GENERAL CONSIDERATIONS CONCERNING THE DRYING OF SEEDS. Internatl. Seed Testing Assoc. Proc. 28: 753–784.

———

1963b. EQUIPMENT OF IMPORTANCE FOR SEED DRYING IN EUROPE. Internatl. Seed Testing Assoc. Proc. 28: 793–814.

———

1972. DRYING AND STORING GRAINS, SEEDS, AND PULSES IN TEMPERATE CLIMATES. Wageningen, Holland, Inst. Storage and Processing Agr. Prod. Pub. 205, 333 pp.

KRISHNASWAMY, P.
1952. STORAGE AND GERMINATION OF MILLET SEEDS. Madras Agron. Jour. 39: 485–490.

KULIK, M. M.
1973. SUSCEPTIBILITY OF STORED VEGETABLE SEEDS TO RAPID INVASION BY ASPERGILLUS AMSTELODAMI AND A. FLAVUS AND EFFECT ON GERMINABILITY. Seed Sci. and Technol. 1: 799–803.

______ and HANLIN, R. T.
1968. OSMOPHILIC STRAINS OF SOME ASPERGILLUS SPP. Mycologia 60: 961–964.

______ and JUSTICE, O. L.
1967. SOME INFLUENCES OF STORAGE FUNGI, TEMPERATURES AND RELATIVE HUMIDITY ON THE GERMINABILITY OF GRASS SEEDS. Jour. Stored Prod. Res. 3: 335–343.

KUPERMAN, F. M.
1950. MECHANICAL INJURY OF SEEDS. Selek. i Semen. 17: 45–48. [In Russian.]

LAKON, G.
1954. DER KEIMWERT DER NACKTEN KARYOPSEN IM SAATGUT VON HAFER UND TIMOTHEE. Saatgutwirtschaft 6: 259–262.

LAMBOU, M. G., CONDON, M. Z., JENSEN, E., and others.
1948. A RAPID METHOD FOR THE MEASUREMENT OF THE INHIBITION OF DETERIORATION IN INTACT SEEDS. Plant Physiol. 23: 84–97.

LARSON, A. H., HARVEY, R. B., and LARSON, J.
1936. LENGTH OF THE DORMANT PERIOD IN CEREAL SEEDS. Jour. Agr. Res. 52: 811–836.

LEGGATT, C. W.
1929–30. CATALASE ACTIVITY AS A MEASURE OF SEED VIABILITY. Sci. Agr. (Ottawa) 10: 73–110.

LEWIS, D.
1953. SEED STORAGE. John Innes Hort. Inst. Ann. Rpt. 43: 14–15.

LIBBY, W. F.
1951. RADIOCARBON DATES, II. Science 114: 291–296.

LIEBERMAN, F. V., DICKE, F. F., and HILLS, O. A.
1961. SOME INSECT PESTS OF IMPORTANT SEED CROPS. U.S. Dept. Agr. Ybk. 1961: 251–258.

LINDSTROM, E. W.
1942. INHERITANCE OF SEED LONGEVITY IN MAIZE INBREDS AND HYBRIDS. Genetics 27: 154.

______
1943. GENETIC RELATIONS OF INBRED LINES OF CORN. Iowa Corn Res. Inst., Rpt. Agr. Res., pt. 2, pp. 41–45.

LINKO, PEKKA, and SOGN, L.
1960. RELATION OF VIABILITY AND STORAGE DETERIORATION TO GLUTAMIC ACID DECARBOXYLASE IN WHEAT. Cereal Chem. 37: 489–499.

LIPMAN, C. B.
1936. NORMAL VIABILITY OF SEEDS AND BACTERIAL SPORES AFTER EXPOSURES TO TEMPERATURES NEAR THE ABSOLUTE ZERO. Plant Physiol. 11: 201–205.

______ and LEWIS, G. N.
1934. TOLERANCE OF LIQUID-AIR TEMPERATURES BY SEEDS OF HIGHER PLANTS FOR SIXTY DAYS. Plant Physiol. 9: 392–394.

LITYNSKI, M., and URBANIAK, Z.
1958. OBSERWACJE NAD WPLYWEM SWIATLA NA NASIONA NIEKTROYCH GATUNKOW ROSLIN W CZASIE ICH PRZECHOWYWANIA. Hodowla Roslin, Aklimatyzacja i Nasiennictwo, Cent. Inst. Inform. Nauk. Technol. i Ekonom. 2: 21–66.

LOCKETT, M. C., and LUYET, B. J.
1951. SURVIVAL OF FROZEN SEEDS OF VARIOUS WATER CONTENTS. Biodynamica 7: 67–76.

LOWIG, E.
1953. BETRAGE ZUR FRAGE DER SAATGUTLAGERUNG. Saatgut-Wirt. 5: 148–150.

———
1963. MODERNE SAATGUTVERPACKUNG. Internatl. Seed Testing Assoc. Proc. 28: 943–979.

LUTE, A. M.
1928. IMPERMEABLE SEED OF ALFALFA. Colo. Expt. Sta. Bul. 326, 30 pp.

LUTHRA, J. C.
1936. ANCIENT WHEAT AND ITS VIABILITY. Current Sci. [India] 4: 489–490.

LYNES, F. F.
1945. POLYPLOIDY IN SUGAR BEETS INDUCED BY STORAGE OF TREATED SEED. Amer. Soc. Agron. Jour. 37: 402–404.

MCALISTER, D. F.
1943. THE EFFECT OF MATURITY ON THE VIABILITY AND LONGEVITY OF THE SEEDS OF WESTERN RANGE AND PASTURE GRASSES. Amer. Soc. Agron. Jour. 35: 442–453.

MCCOLLUM, J. P.
1953. FACTORS AFFECTING COTYLEDONAL CRACKING DURING THE GERMINATION OF BEANS (PHASEOLUS VULGARIS). Plant Physiol. 28 (2): 267–274.

MCILRATH, W. J., ABROL, Y. P., and HEILIGMAN, FRED.
1963. DEHYDRATION OF SEEDS IN INTACT TOMATO FRUITS. Science 142: 1681–1682.

MACKAY, D. B., and TONKIN, J. H. B.
1967. INVESTIGATIONS IN CROP SEED LONGEVITY. I. ANALYSIS OF LONG-TERM EXPERIMENTS WITH SPECIAL REFERENCE TO THE INFLUENCE OF SPECIES, CULTIVAR, PROVENANCE, AND SEASON. Natl. Inst. Agr. Bot. Jour. 11: 209–225.

MCKEE, ROLAND, and MUSIL, A. F.
1948. RELATION OF TEMPERATURE AND MOISTURE TO LONGEVITY OF SEED OF BLUE LUPINE, LUPINUS ANGUSTIFOLIUS, AUSTRIAN WINTER FIELDPEA, PISUM SATIVUM ARVENSE, AND HAIRY VETCH, VICIA VILLOSA. Amer. Soc. Agron. Jour. 40: 459–465.

MCKENZIE, B. A.
1966. SELECTING A GRAIN DRYING METHOD. Purdue Univ. Coop. Ext. Serv. Bul. AE67, 11 pp.

——— FOSTER, G. H., NOYES, R. T., and THOMPSON, R. A.
1968. DRYERATION—BETTER CORN QUALITY WITH HIGH SPEED DRYING. Purdue Coop. Ext. Serv. Bul. AE72, 20 pp.

MCNEAL, X., and YORK, J. O.
1964. CONDITIONING AND STORING GRAIN SORGHUM FOR SEED. Ark. Agr. Expt. Sta. Bul. 687, 16 pp.

MCROSTIE, G. P.
1939. THE THERMAL DEATH POINT OF CORN FROM LOW TEMPERATURES. Sci. Agr. 19: 687–699.

MADSEN, S. B.
1957. INVESTIGATION OF THE INFLUENCE OF SOME STORAGE CONDITIONS ON THE ABILITY OF SEED TO RETAIN ITS GERMINATING CAPACITY. Internatl. Seed Testing Assoc. Proc. 22: 423–446.

———
1962. GERMINATION OF BURIED AND DRY STORED SEEDS III. Internatl. Seed Testing Assoc. Proc. 27: 920–928.

MAES, E., and BAUWEN, M.
1951. INFLUENCE DES RAYONS INFRA-ROUGES SUR LA CONSERVATION DES GERMES DE BLE. Belge, Ecole Off. Meun. Bul. 13 (6): 125–133.

MAMICPIC, N. G., and CALDWELL, W. P.
1963. EFFECTS OF MECHANICAL DAMAGE AND MOISTURE CONTENT UPON VIABILITY OF SOYBEANS IN SEALED STORAGE. Assoc. Off. Seed Anal. Proc. 53: 215–220.

MARTIN, J. A., SENN, T. L., and CRAWFORD, J. H.
1960. RESPONSE OF OKRA SEED TO MOISTURE CONTENT AND STORAGE TEMPERATURE. Amer. Soc. Hort. Sci. Proc. 75: 490–494.

MARTIN, J. N.
1944. CHANGES IN THE GERMINATION OF RED CLOVER SEED IN STORAGE. Iowa Acad. Sci. Proc. 51: 229–233.

______
1945. GERMINATION STUDIES OF SWEET CLOVER SEED. Iowa State Col. Jour. Sci. 19: 289–300.

______ and WATT, J. R.
1944. THE STROPHIOLE AND OTHER SEED STRUCTURES ASSOCIATED WITH HARDNESS IN MELILOTUS ALBA L. AND M. OFFICINALIS WILLD. Iowa State Col. Jour. Sci. 18: 457–469.

MAY, E. N.
1964. HOW TO DRY YOUR SEED STORAGE ROOM. Seedsmen's Digest (Apr.), pp. 26–29.

MEAD, H. W., RUSSELL, R. C., and LEDINGHAM, R. J.
1942. THE EXAMINATION OF CEREAL SEEDS FOR DISEASE AND STUDIES ON THE EMBRYO EXPOSURE IN WHEAT. Sci. Agr. 23: 27–40.

MILES, L. E.
1939. EFFECT OF PERIOD AND TYPE OF STORAGE OF COTTON SEED AFTER TREATMENT WITH ORGANIC MERCURY DUSTS. (Abstract) Phytopathology 29: 754.

______
1941. EFFECT OF STORAGE OF TREATED COTTON SEED IN CLOSELY WOVEN COTTON BAGS. (Abstract) Phytopathology 31: 768–769.

MILLER, E. C.
1938. PLANT PHYSIOLOGY. Ed. 2, 1201 pp. New York.

MILLIGAN, R., and HAYES, P.
1962. LONGEVITY OF LOLIUM SEED UNDER COMMERCIAL STORAGE CONDITIONS. PART II. LONGEVITY OF LOLIUM SEED STORED IN HESSIAN AND IN POLYTHENE LINED HESSIAN SACKS. Internatl. Seed Testing Assoc. Proc. 27: 944–951.

MILNER, M., and GEDDES, W. F.
1946. GRAIN STORAGE STUDIES III. THE RELATION BETWEEN MOISTURE CONTENT, MOLD GROWTH, AND RESPIRATION OF SOYBEANS. Cereal Chem. 23: 225–247.

MIYAGI, K.
1966. EFFECT OF MOISTURE-PROOF PACKING ON THE MAINTENANCE OF VIABILITY OF VEGETABLE SEEDS. Internatl. Seed Testing Assoc. 31: 213–222.

MOODIE, A. W. S.
1925. AGRICULTURAL SEEDS FROM OVERSEAS. EFFECT OF THE VOYAGE ON GERMINATION CAPACITY. Agr. Gaz. N.S. Wales 36: 877–878.

MOORE, R. P.
1955. LIFE ALONE IS NOT ENOUGH—HOW ALIVE ARE SEEDS? Seedmen's Digest 6 (9): 12–13, 37.

MOORE, R. P.

1972. EFFECTS OF MECHANICAL INJURIES ON VIABILITY. *In* Roberts, E. H., Viability of Seeds, pp. 94–113, illus. London.

MOSS, H. J., DERERA, N. F., and BALAAM, L. N.

1972. EFFECT OF PRE-HARVEST RAIN ON GERMINATION IN THE EAR AND α-AMYLASE ACTIVITY OF AUSTRALIAN WHEAT. Austral. Jour. Agr. 23: 769–777.

MUNFORD, R. S.

1965. CONTROLLED TEMPERATURE AND HUMIDITY STORAGE. Miss. State Univ. Short Course Seedsmen Proc. 1965, pp. 144–156.

MUSKETT, A. E.

1948. TECHNIQUE FOR THE EXAMINATION OF SEEDS FOR THE PRESENCE OF SEED-BORNE FUNGI. Brit. Mycol. Soc. Trans. 30: 74–83.

MYERS, AMY.

1942. COLD STORAGE OF ONION SEED. Agr. Gaz. N. S. Wales 53: 528.

MYERS, M. T.

1924. THE INFLUENCE OF BROKEN PERICARP ON THE GERMINATION AND YIELD OF CORN. Amer. Soc. Agron. Jour. 16: 540–552.

NAKAJIMA, Y.

1927. [UNTERSUCHUNGEN UBER DIE KEIMFAHIGKEITSDAUER DER SAMEN.] Bot. Mag. [Tokyo] 41: 604–632. [In Japanese. German summary.]

NAKAMURA, S.

1958. STORAGE OF VEGETABLE SEEDS. Hort. Assoc. Japan, Jour. 27: 32–44.

——— SATO, T., SATO, T., and MINE, T.

1972. STORAGE OF VEGETABLE SEEDS DRESSED WITH FUNGICIDAL DUSTS. Internatl. Seed Testing Assoc. Proc. 37: 961–968.

NAUMOVA, N. A.

1970. ANALIZ SEMYAN NA GRIBNUYA I BAKTERIAL 'NUYA INFEKTIYN. Izadatel 'stvo "Kolos." Ed. 3. Leningrad.

NAVASHIN, M.

1933. ORIGIN OF SPONTANEOUS MUTATIONS. Nature [London] 131: 436.

——— and GERASSIMOWA, H.

1936. NATUR UND URSACHEN DER MUTATIONEN. I. DAS VERHALTEN UND DIE ZYTOLOGIE DER PFLANZEN, DIE AUS INFOLGE ALTERNS MUTIERTEN KEIMEN STAMMEN. Cytologia (Tokyo) 7: 324–362.

——— and SHKVARNIKOV, P.

1933. PROCESS OF MUTATION IN RESTING SEEDS ACCELERATED BY INCREASED TEMPERATURE. Nature 132: 482–483.

NEAL, N. P., and DAVIS, J. R.

1956. SEED VIABILITY OF CORN INBRED LINES AS INFLUENCED BY AGE AND CONDITIONS OF STORAGE. Agron. Jour. 48: 383–384.

NEERGAARD, PAUL.

1973. DETECTION OF SEED-BORNE PATHOGENS BY CULTURE TESTS. Seed Sci. and Technol. 1: 217–254.

NEWHALL, A. G., and HOFF, J. K.

1960. VIABILITY AND VIGOR OF 22 YEAR OLD ONION SEED. Seed World 86 (12): 4–5.

NICHOLS, C., JR.

1941. SPONTANEOUS CHROMOSOME ABERRATIONS IN ALLIUM. Genetics 26: 89–100.

———

1942. THE EFFECTS OF AGE AND IRRADIATION ON CHROMOSOMAL ABERRATIONS IN ALLIUM SEED. Amer. Jour. Bot. 29: 755–759.

NUTILE, G. E.

1958. GERMINATION OF ALFALFA AND RED CLOVER AFTER 24 YEARS OF STORAGE. Assoc. Off. Seed Anal. News Letter 32: 21–22.

______

1964a. DON'T STORE SEEDS NEAR ZERO HUMIDITY. Crops and Soils 16 (4–5): 19–20.

______

1964b. EFFECT OF DESICCATION ON VIABILITY OF SEEDS. Crop Sci. 4: 325–328.

______ and NUTILE, L. C.

1947. EFFECT OF RELATIVE HUMIDITY ON HARD SEEDS IN GARDEN BEANS. Assoc. Off. Seed Anal. Proc. 37: 106–114.

______ and WOODSTOCK, L. W.

1967. THE INFLUENCE OF DORMANCY-INDUCING DESICCATION TREATMENTS ON THE RESPIRATION AND GERMINATION OF SORGHUM. Physiol. Plant. 20: 554–561.

OATHOUT, C. H.

1928. THE VITALITY OF SOYBEAN SEEDS AS AFFECTED BY STORAGE CONDITIONS AND MECHANICAL INJURY. Amer. Soc. Agron. Jour. 20: 837–855.

OHGA, I.

1923. ON THE LONGEVITY OF SEEDS OF NELUMBO NUCIFERA. Bot. Mag. (Tokyo) 37: 87–95.

OWEN, E. G.

1956. THE STORAGE OF SEEDS FOR MAINTENANCE OF VIABILITY. Commonwealth Agr. Bur. Pastures and Field Crops Bul. 43, 81 pp.

OXLEY, T. A.

1948. SCIENTIFIC PRINCIPLES OF GRAIN STORAGE. 103 pp. North. Pub. Co., Liverpool.

PACK, D. A., and OWEN, F. V.

1950. VIABILITY OF SUGAR BEET SEED HELD IN COLD STORAGE FOR 22 YEARS. Amer. Soc. Sugar Beet Technol. Proc. 6: 127–129.

PASSERINI, NAPOLEONE.

1933. SUL PROBABILE SIGNIFICATO BIOLOGICO DEI COSIDDETTI "SEMI DURI." Nuovo Gior. Bot. Ital. 40: 230–251.

PATE, J. B., and DUNCAN, E. N.

1964. VIABILITY OF COTTONSEED AFTER LONG PERIODS OF STORAGE. Crop Sci. 4: 342–344.

PATRICK, S. R.

1936. SOME OBSERVATIONS ON THE DECLINE IN VIABILITY OF CHEWINGS FESCUE SEED (FESTUCA RUBRA COMMUTATA). Assoc. Off. Seed Anal. Proc. 28: 76–79.

PERRY, J. S., and HALL, C. W.

1960. GERMINATION OF PEA BEANS AS AFFECTED BY MOISTURE AND TEMPERATURE AT IMPACT LOADING. Mich. Agr. Expt. Sta. Quart. Bul. 43: 33–39.

PETER, A.

1893. KULTUREVERSUCHE MIT RUHENDEN SAMEN. I. Nachr. Göttingen Gesell. Wiss. 1893: 673–691.

PETERSON, ANNE, SCHLEGEL, VERA, HUMMEL, B., and others.

1956. GRAIN STORAGE STUDIES. XXII. INFLUENCE OF OXYGEN AND CARBON DIOXIDE CONCENTRATIONS ON MOLD GROWTH AND GRAIN DETERIORATION. Cereal Chem. 33: 53–66.

PETERSON, W. H.

1971. IN STORAGE DRYING WITH SUPPLEMENTAL HEAT. S. Dak. State Univ. Ext. Bul. FS531, 6 pp.

PETO, F. H.

1933. THE EFFECT OF AGING AND HEAT ON THE CHROMOSOMAL MUTATION RATES IN MAIZE AND BARLEY. Canad. Jour. Res. 9: 261–264.

PHILLIS, E., and MASON, T. G.

1945. THE EFFECT OF EXTREME DESICCATION ON THE VIABILITY OF COTTON SEED. Ann. Bot. 9: 353–359.

PHILPOT, RAY.

1976. PRINCIPLES AND PRACTICES OF SEED DRYING. Miss. State Univ. Short Course Seedsmen Proc. 16 (1973): 23–40.

PIXTON, S. W., and WARBURTON, S.

1968. THE TIME REQUIRED FOR CONDITIONING GRAIN TO EQUILIBRIUM WITH SPECIFIC RELATIVE HUMIDITIES. Stored Prod. Res. Jour. 4: 261–265.

POLLOCK, B. M.

1961. THE EFFECTS OF PRODUCTION PRACTICES ON SEED QUALITY. Seed World 89 (5): 6, 8, 10.

——— and MANALO, J. R.

1970. SIMULATED MECHANICAL DAMAGE TO GARDEN BEANS DURING GERMINATION. Amer. Soc. Hort. Sci. Jour. 95 (4): 415–417.

——— ROOS, E. E., and MANALO, J. R.

1969. VIGOR OF GARDEN BEAN SEEDS AND SEEDLINGS INFLUENCED BY INITIAL SEED MOISTURE, SUBSTRATE, OXYGEN, AND INBIBITION TEMPERATURE. Amer. Soc. Hort. Sci. Jour. 94 (6): 577–584.

POMERANZ, Y.

1966. THE ROLE OF THE LIPID FRACTION IN GROWTH OF CEREALS, AND IN THEIR STORAGE AND PROCESSING. Wallerstein Lab. Commun. 29 (98/99): 17–26.

PORSILD, A. E., HARINGTON, C. R., and MULLIGAN, G. A.

1967. LUPINUS ARCTICUS WATS. GROWN FROM SEEDS OF PLEISTOCENE AGE. Science 158: 113–114.

POTGIETER, H. V., and BEER, P. R. DE.

1972. THE EFFECT OF CERTAIN CONTACT INSECTICIDES AND FUMIGANTS ON THE GERMINATION, AFTERGROWTH AND YIELD OF STORED GRAIN SEED. Internatl. Seed Testing Assoc. Proc. 37: 941–951.

PRITCHARD, E. W.

1933. HOW LONG DO SEEDS RETAIN THEIR GERMINATING POWER? Jour. Dept. Agr. N.S. Wales 36: 645–646.

QASEM, S. A., and CHRISTENSEN, C. M.

1958. INFLUENCE OF MOISTURE CONTENT, TEMPERATURE, AND TIME ON THE DETERIORATION OF STORED CORN BY FUNGI. Phytopathology 48: 544–549.

RADKO, A. M.

1955. ZOYSIA SEED STORAGE AND GERMINATION TESTS. U.S. Golf Assoc. Jour. 8 (6): 25–26.

REES, BERTHA.

1911. LONGEVITY OF SEEDS AND STRUCTURE AND NATURE OF SEED COAT. Roy. Soc. Victoria, Proc. (N.S.) 23: 393–414.

REEVES, R. G.

1936. COMPARATIVE ANATOMY OF THE SEEDS OF COTTONS AND OTHER MALVACEOUS PLANTS. Amer. Jour. Bot. 23: 291–296, 394–405.

RIDDELL, J. A., and GRIES, G. A.

1956. THE INFLUENCE OF AGE AND MATURATION TEMPERATURE OF WHEAT GRAINS ON PLANT DEVELOPMENT. Ind. Acad. Sci. Proc. 66: 62.

ROBBINS, W. A., and PORTER, R. H.

1946. GERMINABILITY OF SORGHUM AND SOYBEAN SEED EXPOSED TO LOW TEMPERATURES. Amer. Soc. Agron. Jour. 38: 905–913.

ROBERTS, E. H.
1972. VIABILITY OF SEEDS. 448 pp., illus. London.
_____ and ABDALLA, F. H.
1968. THE INFLUENCE OF TEMPERATURE, MOISTURE, AND OXYGEN ON PERIOD OF SEED VIABILITY IN BARLEY, BROAD BEANS, AND PEAS. Ann. Bot. (N.S.) 32: 97–117.
ROBERTS, H. M.
1959. THE EFFECT OF STORAGE CONDITIONS ON THE VIABILITY OF GRASS SEEDS. Internatl. Seed Testing Assoc. Proc. 24: 184–213.
_____
1962. THE EFFECT OF STORAGE CONDITIONS ON THE VIABILITY OF TIMOTHY SEEDS. Welsh Plant Breeding Sta. Rpt. 1961: 99.
ROBERTSON, D. W., LUTE, A. M., and KROEGER, H.
1943. GERMINATION OF 20-YEAR-OLD WHEAT, OATS, BARLEY, CORN, RYE, SORGHUM, AND SOYBEANS. Amer. Soc. Agron. Jour. 35: 786–795.
RODRIGO, P. A.
1935. LONGEVITY OF SOME FARM CROP SEEDS. Philippine Jour. Agr. 6: 343–357.
_____
1939. STUDY ON THE VITALITY OF OLD AND NEW SEEDS OF MUNGO (PHASEOLUS AUREUA ROXB.). Philippine Jour. Agr. 10: 285–291.
_____
1953. SOME STUDIES ON THE STORING OF TROPICAL AND TEMPERATE SEEDS IN THE PHILIPPINES. 13th Internatl. Hort. Cong. Rpt. 2: 1061–1066.
ROSSMAN, E. C.
1949. FREEZING INJURY OF MAIZE SEED. Plant Physiol. 24: 629–656.
RUAN, E., and FREY, K. J.
1957. EFFECT OF HEAT TREATMENT ON OAT SEEDS. Iowa Acad. Sci. Proc. 64: 139–148.
RUSSELL, R. C.
1958. LONGEVITY STUDIES WITH WHEAT SEED AND CERTAIN SEED-BORNE FUNGI. Canad. Jour. Plant Sci. 38: 29–33.
SAHADEVAN, P. C., and RAO, M. B. V. N.
1947. NOTE ON THE DETERIORATION IN GERMINATION CAPACITY OF A PADDY STRAIN IN MALABAR AND SOUTH KANARA. Current Sci. (India) 16: 319–320.
SAMPIETRO, G.
1931. PROLONGING THE LONGEVITY OF RICE SEED. Gior. di Risic. 21: 1–5.
SAN PEDRO, A. V.
1936. INFLUENCE OF TEMPERATURE AND MOISTURE ON THE VIABILITY OF VEGETABLE SEEDS. Philippine Agr. 24: 649–658.
SAYRE, J. D.
1940. STORAGE TESTS WITH SEED CORN. Ohio Jour. Sci. 40: 181–185.
SCHJELDERUP-EBBE, T.
1936. UBER DIE LEBENSFAHIGKEIT ALTER SAMEN. Norske Vidensk. Akad. i Oslo Math. Nat. Kl. Skr. 1935, No. 13, 178 pp.
SCHLOESING, A. T., and LEROUX, D.
1943. ESSAI DE CONSERVATION DE GRAINES EN L'ABSENCE D'HUMIDITE, D'AIR ET DE LUMIERE. [STUDIES ON THE CONSERVATION OF SEEDS IN THE ABSENCE OF MOISTURE, AIR, AND LIGHT.] Acad. Agr. Compt. Rend. 28–29; 204–206.
SCHROEDER, H. W., and ROSBERG, D. W.
1959. DRYING ROUGH RICE WITH INFRA-RED RADIATION. Tex. Agr. Ext. Serv. MP 354, 4 pp.
SCOTT, D. H., and DRAPER, A. D.
1970. A FURTHER NOTE ON LONGEVITY OF STRAWBERRY SEEDS IN COLD STORAGE. HortScience 5: 439.

SEMENIUK, G.
1954. MICROFLORA. *In* Anderson, J. A., and Alcock, A. W., eds., Storage of Cereal Grains and Their Products, pp. 77–151, illus. St . Paul, Minn.

SENGBUSCH, R. VON.
1955. DIE ERHALTUNG DER KEIMFAHIGKEIT VON SAMEN BEI TIEFEN TEMPERATUREN. Züchter 25: 168–169.

SHANDS, H. L., JANISCH, D. C., and DICKSON, A. D.
1967. GERMINATION RESPONSE OF BARLEY FOLLOWING DIFFERENT HARVESTING CONDITIONS AND STORAGE TREATMENTS. Crop Sci. 7: 444–446.

SIFTON, H. B.
1920. LONGEVITY OF THE SEEDS OF CEREALS, CLOVERS, AND TIMOTHY. Amer. Jour. Bot. 7: 243–251.

SIJBRING, P. H.
1963. CONDITIONING OF ROOMS FOR SEED STORAGE. Internatl. Seed Testing Assoc. Proc. 28: 885–892.

SIMPSON, D. M.
1942. FACTORS AFFECTING THE LONGEVITY OF COTTON SEED. Jour. Agr. Res. 64: 407–419.

———
1946. THE LONGEVITY OF COTTONSEED AS AFFECTED BY CLIMATE AND SEED TREATMENTS. Amer. Soc. Agron. Jour. 38: 32–45.

———
1953. COTTONSEED STORAGE IN VARIOUS GASES UNDER CONTROLLED TEMPERATURE AND MOISTURE. Tenn. Agr. Expt. Sta. Bul. 228, 16 pp.

——— ADAMS, C. L., and STONE, G. M.
1940. ANATOMICAL STRUCTURE OF THE COTTONSEED COAT AS RELATED TO PROBLEMS OF GERMINATION. U.S. Dept. Agr. Tech. Bul. 734, 24 pp.

——— and MILLER, P. R.
1944. THE RELATION OF ATMOSPHERIC HUMIDITY TO MOISTURE IN COTTONSEED. Amer. Soc. Agron. Jour. 36: 957–959.

SINHA, R. N., and WALLACE, H. A. H.
1965. ECOLOGY OF A FUNGUS-INDUCED HOT SPOT IN STORED GRAIN. Canad. Jour. Plant Sci. 45: 48–59.

SIVORI, E., NAKAYAMA, F., and CIGLIANO, E.
1968. GERMINATION OF ACHIRA SEED (CANNA SP.) APPROXIMATELY 550 YEARS OLD. Nature 219: 1269–1270.

SMITH, A. J. M., and GANE, R.
1938. THE WATER RELATIONS OF SEEDS. *In* (Brit.) Dept. Sci. and Indus. Res., Report of the Food Investigation Board for the Year 1938, pp. 231–240. London.

SNOW, D., CRICHTON, M. H. G., and WRIGHT, N. C.
1944. MOULD DETERIORATION OF FEEDING-STUFFS IN RELATION TO HUMIDITY OF STORAGE. PART I. THE GROWTH OF MOULDS AT LOW HUMIDITIES. Ann. Appl. Biol. 31: 102–110.

SPENCER, D. A.
1954. RODENTS. *In* Anderson, J. A., and Alcock, A. W., eds., Storage of Cereal Grains and Their Products, pp. 275–307, illus. St. Paul, Minn.

STEVENS, O. A.
1935. GERMINATION STUDIES ON AGED AND INJURED SEEDS. Jour. Agr. Res. 51: 1093–1106.

——— and LONG, H. D.
1926. SWEET CLOVER SEED STUDIES. N. Dak. Agr. Expt. Sta. Bul. 197, 20 pp.

STEVENSON, F. J., and EDMUNDSON, W. C.
1950. STORAGE OF POTATO SEEDS. Amer. Potato Jour. 27: 408–411.

STREETER, J. G.
1965. POSSIBLE MECHANISMS IN THE LOSS OF SEED VIABILITY. Assoc. Off. Seed Anal. Newsletter 39: 27–35.

TAMMES, T.
1900. UEBER DEN EINFLUSS DER SOMENSTRAHLEN AUF DIE KEIMUNGSFACHIGKEIT VON SAMEN. Landw. Forsch. 29: 467–482.

THOMPSON, H. J., and SHEDD, C. K.
1954. EQUILIBRIUM MOISTURE AND HEAT OF VAPORIZATION OF SHELLED CORN AND WHEAT. Agr. Engin. 35: 786–788.

THRONEBERRY, G. O., and SMITH, F. G.
1955. RELATION OF RESPIRATORY AND ENZYMATIC ACTIVITY TO CORN SEED VIABILITY. Plant Physiol. 30: 337–343.

TOOLE, E. H.
1942. STORAGE OF VEGETABLE SEEDS. U.S. Dept. Agr. Leaflet 220, 8 pp.

______
1950. RELATION OF SEED PROCESSING AND OF CONDITIONS DURING STORAGE ON SEED GERMINATION. Internatl. Seed Testing Assoc. Proc. 16: 214–225.

______ and BROWN, E.
1946. FINAL RESULTS OF THE DUVEL BURIED SEED EXPERIMENT. Jour. Agr. Res. 72: 201–210.

______ and TOOLE, V. K.
1946. RELATION OF TEMPERATURE AND SEED MOISTURE TO THE VIABILITY OF STORED SOYBEAN SEED. U.S. Dept. Agr. Cir. 753, 9 pp.

______ and TOOLE, V. K.
1954. RELATION OF STORAGE CONDITIONS TO GERMINATION AND TO ABNORMAL SEEDLINGS OF BEAN. Internatl. Seed Testing Assoc. Proc. 18: 123–129.

______ and TOOLE, V. K.
1960. VIABILITY OF SNAP BEAN SEED AS AFFECTED BY THRESHING AND PROCESSING INJURY. U.S. Dept. Agr. Tech. Bul. 1213, 9 pp.

______ TOOLE, V. K., and GORMAN, E. A.
1948. VEGETABLE-SEED STORAGE AS AFFECTED BY TEMPERATURE AND RELATIVE HUMIDITY. U.S. Dept. Agr. Tech. Bul. 972, 24 pp.

______ TOOLE, V. K., LAY, B. J., and CROWDER, J. T.
1951. INJURY TO SEED BEANS DURING THRESHING AND PROCESSING. U.S. Dept. Agr. Cir. 874, 10 pp.

______ TOOLE, V. K., and NELSON, E. G.
1960. PRESERVATION OF HEMP AND KENAF SEED. U.S. Dept. Agr. Tech. Bul. 1215, 16 pp.

______ TOOLE, V. K., and NELSON, E. G.
1961. PLASTIC BAGS FOR SHIPPING SEEDS IN THE TROPICS. Internatl. Seed Testing Assoc. Proc. 26: 86–88.

TOOLE, V. K.
1939. NOTES ON THE VIABILITY OF THE IMPERMEABLE SEEDS OF VICIA VILLOSA, HAIRY VETCH. Assoc. Off. Seed Anal. 31: 109–111.

______ and TOOLE, E. H.
1953. SEED DORMANCY IN RELATION TO SEED LONGEVITY. Internatl. Seed Testing Assoc. Proc. 18: 325–328.

TUITE, J. F., and CHRISTENSEN, C. M.
1957. GRAIN STORAGE STUDIES. XXIV. MOISTURE CONTENT OF WHEAT SEED IN RELATION TO INVASION OF THE SEED BY SPECIES OF THE ASPERGILLUS GLAUCUS GROUP, AND THE EFFECT OF INVASION UPON GERMINATION OF THE SEED. Phytopathology 47: 323–327.

TURNER, J. H.
1933. THE VIABILITY OF SEEDS. Kew Roy. Bot. Gard. Bul. Misc. Inform. 6: 257–269.

——— and FERGUSON, D.
1972. FIELD PERFORMANCE OF COTTON GROWN FROM FILLED AND PARTIALLY FILLED SEEDS. Crop Sci. 12: 868–871.

ULLMANN, W.
1949. UBER DIE KEIMFAHIGKEITDAUER (LEBENSDAUER) VON LANDWIRTSCHAFTLICHEN UND GARTENBAULICHEN SAMEN. Saatgut-Wirt. 1: 174–175, 195–196.

U.S. DEPARTMENT OF AGRICULTURE.
1968. RULES AND REGULATIONS OF THE SECRETARY OF AGRICULTURE UNDER THE FEDERAL SEED ACT OF AUGUST 9, 1939 (53 STAT. 1275). 83 pp. U.S. Dept. Agr. Consum. and Mktg. Serv., Washington, D.C.

VIBAR, T., and RODRIGO, P. A.
1929. STORING FARM CROP SEEDS. Philippine Agr. Rev. 22: 135–146.

VILLIERS, T. A.
1973. AGEING AND THE LONGEVITY OF SEEDS IN FIELD CONDITIONS. *In* Heydecker, W., Seed Ecology, pp. 265–288. Pa. State Univ. Press, University Park, Pa., and London.

VON DEGEN, A., and PUTTEMANS, A.
1931. SUR L'INFLUENCE DU TRANSPORT MARITIME SUR LA GERMINATION DES SEMENCES. Internatl. Seed Testing Assoc. Proc. 3: 287–291.

WAGGONER, H. D.
1917. THE VIABILITY OF RADISH SEEDS (RAPHANUS SATIVUS L.) AS AFFECTED BY HIGH TEMPERATURES AND WATER CONTENT. Amer. Jour. Bot. 4: 299–313.

WALLACE, H. A. H.
1973. FUNGI AND OTHER ORGANISMS ASSOCIATED WITH STORED GRAIN. *In* Sinha, R. H., and Muir, W. E., Grain Storage: Part of A System, pp. 71–98. Westport, Conn.

——— and SINHA, R. N.
1962. FUNGI ASSOCIATED WITH HOT SPOTS IN FARM STORED GRAINS. Canad. Jour. Plant Sci. 42: 130–141.

WATERS, E. C., JR.
1962. THIRTY-EIGHT YEARS BEHIND THE TIMES AND THEY STILL GERMINATE. Assoc. Off. Seed Anal. Newsletter 36: 8.

WATSON, D. P.
1948. STRUCTURE OF THE TESTA AND ITS RELATION TO GERMINATION IN THE PAPILIONACEAE TRIBES TRIFOLAI AND LOTEAE. Ann. Bot. (N.S.) 12: 385–409.

WEIBULL, G.
1952. THE COLD STORAGE OF VEGETABLE SEED AND ITS SIGNIFICANCE FOR PLANT BREEDING AND THE SEED TRADE. Agr. Hort. Genet. 10: 97–104.

———
1955. THE COLD STORAGE OF VEGETABLE SEED—FURTHER STUDIES. Agr. Hort. Genet. 13: 121–142.

WEISS, M. G., and WENTZ, J. B.
1937. EFFECT OF LUTEUS GENES ON LONGEVITY OF SEED IN MAIZE. Amer. Soc. Agron. Jour. 29: 63–75.

WELCH, G. B.
1967. SEED STORAGE TO MAINTAIN QUALITY. Miss. State Univ. Short Course Seedsmen 1967, pp. 53–63.

WELTON, F. A.
1921. LONGEVITY OF SEEDS. Ohio Agr. Expt. Sta. Monthly Bul. 6 (1-2): 18–25.

WENT, F. W.
1948. AN EXPERIMENT THAT MAY LAST 360 YEARS. PROJECT WILL GIVE BOTANISTS NEW INFORMATION ABOUT THE HEREDITY OF PLANTS AND THE LIFE SPAN OF SEEDS. Life 24 (5): 57–58, 60.

_____ and MUNZ, P. A.
1949. A LONG TERM TEST OF SEED LONGEVITY. El Aliso 2: 63–75.

WESTER, H. V.
1973. FURTHER EVIDENCE OF AGE OF ANCIENT VIABLE LOTUS SEEDS FROM PULANTIEN DEPOSIT, MANCHURIA. HortScience 8: 371–377.

WHEELER, W. A., and HILL, D. D.
1957. GRASSLAND SEEDS. 734 pp., illus. Princeton, N.J.

WHITCOMB, W. O.
1931. HARD SEEDS IN LEGUMES. INTERPRETATION OF THEIR VALUE AND METHODS OF TREATMENT. Mont. Agr. Expt. Sta. Bul. 248, 63 pp.

WHITE, JEAN.
1909. THE FERMENTS AND LATENT LIFE OF RESTING SEEDS. Roy. Soc. London, Proc. 81 (B550): 417–442.

WHITE, O. E.
1946. THE GERMINATION OF PEAS IN FLORIDA AND KING TUT'S TOMB. Turtox News 24: 6–8.

WHITEHEAD, E. I., and GASTLER, G. F.
1946–47. HYGROSCOPIC MOISTURE OF GRAIN SORGHUM AND WHEAT AS INFLUENCED BY TEMPERATURE AND HUMIDITY. S. Dak. Acad. Sci. Proc. 26: 80–84.

WHYMPER, R., and BRADLEY, A.
1947. A NOTE ON THE VIABILITY OF WHEAT SEEDS. Cereal Chem. 24: 228–229.

WILEMAN, R. H., and ULLSTRUP, A. J.
1945. A STUDY OF FACTORS DETERMINING SAFE DRYING TEMPERATURES FOR SEED CORN. Purdue Univ. Agr. Expt. Sta. Bul. 509, 10 pp.

WILLIAMS, M.
1938. THE MOISTURE CONTENT OF GRASS SEED IN RELATION TO DRYING AND STORING. Welsh Jour. Agr. 14: 213–233.

WINCHESTER, W. J.
1954. STORING SEED OF GREEN PANIC AND BUFFEL GRASS FOR BETTER GERMINATION. Queensland Agr. Jour. 79: 203–204.

WOLLENWEBER, H. W.
1942. UBER DIE LEBENSDAUER VON KARTOFFELSAMEN. Angew. Bot. 24: 259–260.

WOODSTOCK, L. W.
1973. PHYSIOLOGICAL AND BIOCHEMICAL TESTS FOR SEED VIGOR. Seed Sci. and Technol. 1: 127–157.

WORTMAN, L. S., and RINKE, E. H.
1951. SEED CORN INJURY AT VARIOUS STAGES OF PROCESSING AND ITS EFFECT UPON COLD TEST PERFORMANCE. Agron. Jour. 43: 299–305.

WRIGHT, W. G., and KINCH, R. C.
1962. DORMANCY IN SORGHUM VULGARE PERS. Assoc. Off. Seed Anal. Proc. 52: 169–177.

WYTTENBACH, E.
1955. DER EINFLUSS VERSCHIEDENER LAGERUNGSFAKTOREN AUF DIE HALTBARKEIT VON FELDSAMEREIEN (LUZERNE, ROTKLEE UND GEMEINEM SCHOTENKLEE) BEI LANGER DAUERNDER AUFBEWAHRUNG. Landw. Jahrb. der Schweiz 69: 161–196.

YOUNG, H. D.

1929. EFFECT OF VARIOUS FUMIGANTS ON THE GERMINATION OF SEEDS. Jour. Agr. Res. 39: 925–927.

YOUNG, R. E.

1949. THE EFFECT OF MATURITY AND STORAGE ON GERMINATION OF BUTTERNUT SQUASH SEED. Amer. Soc. Hort. Sci. Proc. 53: 345–346.

YOUNGMAN, B. J.

1952. GERMINATION OF OLD SEEDS. Kew Bul. 6, pp. 423–426.

ZELENY, LAWRENCE.

1954. CHEMICAL, PHYSICAL, AND NUTRITIVE CHANGES DURING STORAGE. *In* Anderson, J. A., and Alcock, A. W., eds., Storage of Cereal Grains and Their Products, pp. 46–76, illus. St. Paul, Minn.

_____

1961. WAYS TO TEST SEEDS FOR MOISTURE. U.S. Dept. Agr. Ybk. 1961: 443–447.

_____ and COLEMAN, D. A.

1938. ACIDITY IN CEREALS AND CEREAL PRODUCTS, ITS DETERMINATION AND SIGNIFICANCE. Cereal Chem. 15: 580–595.

_____ and COLEMAN, D. A.

1939. THE CHEMICAL DETERMINATION OF SOUNDNESS IN CORN. U.S. Dept. Agr. Tech. Bul. 644, 24 pp.

INDEX